MIKE,

On the occasion of your 60th birthday, may all your peaches be large, sweet, and plentiful!

with much love,

KATHY & SCOTT

June 2012

Texas Peach Handbook

AgriLife **EXTENSION**
Texas A&M System

AgriLife Research and Extension Service
Series

Craig Nessler and Edward G. Smith
General Editors

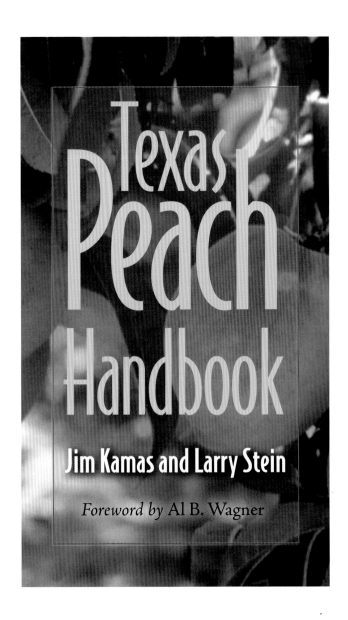

Texas
Peach
Handbook

Jim Kamas and Larry Stein

Foreword by Al B. Wagner

TEXAS A&M UNIVERSITY PRESS

College Station

LIBRARY OF CONGRESS CATALOGING-IN-PUBLICATION DATA

Kamas, Jim, 1955–
Texas peach handbook / Jim Kamas and Larry Stein;
foreword by Al B. Wagner.—1st ed.
p. cm.—(AgriLife Research and Extension Service series)
Includes index.
ISBN 978-1-60344-266-4 (flexibound: alk. paper)
1. Peach—Texas.   I. Stein, Larry, 1956–   II. Title.
III. Series: AgriLife Research and Extension Service series.
SB371.K15   2011
634.2509764—dc22
2010035647

*This book
is dedicated to our teachers
and to future peach growing enthusiasts
of the South.*

Contents

· ·

Foreword

· ·

It is my privilege to provide some opening words for this comprehensive publication on growing peaches. For years growers have relied on the *Peach Handbook* for information on peach production. *Texas Peach Handbook* combines all the information from that handbook with more current technologies to consider when cultivating peaches. It will be useful to commercial growers as well as to homeowners with one or two trees in their backyard. Jim Kamas and Larry Stein have more than fifty-five combined years of experience with peach production. I have known Jim and Larry for almost thirty years; twelve of those years I was fortunate to serve as their program leader. During that time, I had many opportunities to observe them speaking to groups and working with growers one-on-one. Believe them—they are two of the most dedicated and respected horticulturists in the country.

It is my opinion that you will find everything you need to know about growing peaches in this book. It is all here, from site selection and pest management to harvesting and marketing. There is nothing better than a ripe peach picked from your tree, and this publication will guide you to that ultimate and delicious goal.

—Al B. Wagner
　　Professor and AgriLife Extension Food Technologist
　　Texas A&M University System

Preface

· ·

With more than thirty years of research and extension work in peach orchards, the authors have come to realize the need for a practical handbook to educate both commercial and home peach growers. While *Texas Peach Handbook* is by no means a comprehensive review of all available work on peach growing, it will serve as a strong foundational guide for amateur and beginning peach growers as well as offer new and improved growing practices for the experienced grower. Through written with Texas conditions in mind, it should prove to be a reliable reference to peach growers across the south and Gulf Coast. The authors seek not only to inform readers of how to successfully establish and maintain peach orchards, but also to shed light on the scientific principles behind our decisions. Understanding these principles will maximize the enjoyment and profitability of growing peaches and provide the best opportunity for a healthy and long-lived planting. But, there is indeed an art to understanding the nuanced differences in soils, climate, and individual tree performance that are unique to each orchard site. Each year offers new lessons to learn; no two growing seasons are alike. The old adage that the greatest fertilizer in an orchard is the grower's shadow is true. The more time a grower spends walking the orchard, the quicker observational skills will be honed along with the ability to anticipate needs or come up with solutions in a timely manner.

Acknowledgments

· ·

This book has only been possible because of the work and wisdom of our teachers and mentors throughout our careers. We owe our collective knowledge and perspective to numerous scholars, but our horticultural mentors have had the greatest impact on our lives, namely J. Benton Storey, Hollis Bowen, Bluefford Hancock, and George Ray McEachern. They inspired us to search for more than answers and to seek long-range solutions to problems. Due to their efforts we are able to provide a simple but technically accurate publication on growing peaches. We must also acknowledge the work of other extension horticulturists, their knowledge, and their photographs. Over the years we have all accumulated photos that have become the heart of our educational program, and we thank all concerned for their efforts and the use of their photos, as numerous such "extension photos" appear in this book.

Texas Peach Handbook

Planning a Peach Orchard

hen contemplating planting a peach orchard, it is important first to address the question of what one is going to do with the fruit. This may sound obvious, but in many ways, growing the fruit can be the easy part. More than a few growers have struggled to establish an orchard and produce a first crop, only to find themselves with ripe fruit and no market for their crop. Peaches are extremely perishable, and marketing arrangements should be made far in advance of your first harvest. The location of a wholesale operation can be somewhat more flexible, but when contemplating a pick-your-own or roadside sale orchard, proximity to a population center or major thoroughfare is critical to sales. It is our opinion that no peach tree should be planted without first knowing where and how the fruit will be marketed.

One of the more common mistakes made by novice growers is to underestimate the time required to manage peach trees properly. Although mechanical aids are available for pruning and thinning, no more than 5 acres should be planted as a part-time enterprise unless a ready supply of supplemental labor can be hired as needed. In any case, a person planning a part-time peach enterprise should plan on having most weekends and many evenings tied up from mid-January through the harvest season. Bad weather on weekends can necessitate scheduling sprays during the week and may require taking vacation time. Unfortunately, 5-acre peach operations require almost as much equipment as 20-acre plantings, and production costs may therefore be higher for the part-time grower. One person should not attempt more than 20 acres as a full-time occupation unless additional help is available. Even in this scenario, additional help may be needed for pruning mature trees and will certainly be required for fruit thinning and harvest. The time factor, particularly for a part-time grower, must not be ignored. Even though it may appear that good profit potential exists, each person must place his or her own value on family and leisure time and make decisions accordingly.

The decision of whether to go into peach production should be carefully considered, and all factors should be studied before a final decision is made. This book is designed to assist potential peach growers in deciding whether to enter the orchard business and will serve as a beginner's guide for orchard operations. Planning the planting of a new orchard should begin at least one year in advance so that unrushed, intelligent decisions can be made and carried out concerning site selection and preparation, any potential equipment needs, variety and rootstock selection, tree planting, and weed control issues.

For personal and small supplemental income orchards, growers commonly plant either at their home or on property they already own close to home. This may make sense, but when a significant invest-

High quality peach ready for harvest.

ment with larger plantings is involved, prospective growers should resist the urge to plant the "old back forty" just because they own it. Site selection can and will make or break a commercial planting. A site that is prone to spring frost will be at significantly greater risk of losing a crop year after year, and as the old timers say, "they don't pay you for growing leaves." Even on the best sites, peach growers in Texas normally expect to lose a crop completely to spring frost about one in seven years.

Planting in frost-prone areas can increase the risk of total crop loss severalfold, to the point that the orchard is simply financially unsustainable. Consider your relationship with your orchard to be like a marriage. Choosing the wrong site may disrupt that marriage and can certainly impact other interpersonal relationships as well. Losing a large investment as a result of poor site selection will not make you a happy person, and this will no doubt impact all other aspects of your life.

After selecting a site, the proper selection of rootstocks and varieties becomes the second most important decision ahead of a prospective grower. Again, do your homework, talk to growers in adjacent regions, consult the agricultural extension people in your county, and start talking to commercial peach tree nurseries early in the planning process. Avoid the temptation to plant what the nurseries happen to have on hand at the time you order. In an ideal world, a new grower should be communicating with the nursery of choice at least one year prior to the anticipated date of planting.

Choose varieties with the appropriate chilling requirement for your area and that have fruit characteristics that fit your marketing scheme. For example, a variety such as 'Redskin' may be consistently productive and have great fruit quality but may have somewhat of a soft suture at maturity. If a grower plans on selling fruit in a roadside stand or as a pick-your-own fruit operation, this variety may be a good choice. If the marketing scheme calls for large-scale harvest and shipping long distances, the same variety may prove disastrous. Don't rush these decisions. Plan your operation with a clear vision of what you intend to do with your crop.

Once trees are planted, irrigation, floor management (weed control), and nutrition are the keys to growing and maintaining a strong orchard. When one adds nitrogen and water to a site in the southern

Quality fruit can indeed be grown by homeowners and commercial growers alike.

United States, weed populations explode. Remember, the trees you are planting are a non-native, introduced crop. The weeds were here first and will win any contest for water, nutrients, or sunlight. Failure to control weeds adequately is the number one cause of failure in new orchards. If you choose to plant an orchard, you will spend a larger percentage of your time fighting weeds than you ever imagined. Be smart, be diligent, and be forewarned.

Peach production, like that of many other perennial crops, can provide substantial challenges in insect and disease control. One of the major differences between growing annual and perennial crops is that when a serious disease outbreak drastically affects an annual crop, the grower has the option to rotate to a different crop the following year. Obviously this is not the case with perennial crops. As orchardists, we are challenged with the management of insects and diseases on a given crop at a given location for decades.

The key to proper management is to "know thy enemies" fully. Educate yourself on all the pitfalls that may potentially threaten your trees or your crop. Understand how cultural practices affect insect and disease pressure, and make management decisions accordingly. On a commercial scale, peaches can be considered to be a high-input crop. In other words, organic peach production in our climate is very, very risky. We can mitigate risk with cultural practices, but in some years, at most locations, chemical inputs will be needed if an orchard is to produce an economically sustainable crop of commercially acceptable fruit.

In addition to choosing an appropriate site, how we manage our trees can drastically impact their ability to withstand late spring frosts and winter freezes. In the section about this in chapter 4 we attempt to explain enough about how and why trees respond to the environment to have our recommendations make sense without taking you through a semester of plant physiology. Growing productive, healthy and long-lived orchards is not rocket science, but neither is it as elementary as many believe. In many ways rocket science may be less challenging. Successful orchardists are well versed in both the science and the art of pomology and must be dedicated to a career of continual learning. No two sites or seasons are alike; growers must always be alert and be flexible in their decision making. One must be able to learn from current problems and react appropriately to remedy them in the future. Inappropriate decisions can negate years of previous effort and can impact crops years into the future.

As mentioned, peaches are a very perishable commodity. With local sales, only limited sorting, washing, and storage facilities are needed, but for larger operations, proper storage is essential for maintaining fruit quality until the peaches can be delivered to market. Understanding the ripening process and how the environment affects fruit quality and shelf life are critical components of providing your customers with the highest fruit quality available. Remember, there will always be cheaper fruit available at wholesale markets or even at grocery outlets. What you have to offer, as a local grower of tree-ripened peaches, is high quality fruit. You should start your orchard venture with this in mind. You must be prepared to produce the highest quality fruit available because you will never be able to compete with these markets on price. A great majority of our consuming public is willing to pay a premium for this measure of increased fruit quality. We have all been disappointed time and time again by purchasing peaches that are attractive and blemish free, coming from distant growing locations, only to have them taste just slightly better than the box. Quality sells and is its own best advertisement.

Last but certainly not least, marketing and economics play a major role in the economic sustainability of a peach orchard. While they can be appealing at first glance, pick-your-own enterprises can be economically devastating when weather conditions do not favor day-trippers coming to your location, and dealing with the general public can be at times challenging. For this type of operation, you need to be a people person as well as a peach grower. What must be accomplished in the pick-your-own operation is the marketing of "quality of life" experiences. The opportunities for hands-on experiences with our food are increasingly rare in our culture and can command a premium.

Large-scale commercial growers can find themselves in a pinch when wholesale prices plummet as a result of imported fruit. For smaller-scale fruit growers, the economic risks may be fewer, but everyone growing peaches assumes risk both personal and economic. There are many, many easier ways to make an income than growing peaches, but for a rare few, we would simply not be personally satisfied if this were not what we did with our lives.

For planning purposes, here are some general rules of thumb:

- The standard spacing for peach trees is approximately 20 X 20 feet, which is approximately 100 trees per acre. To allow for equipment movement through the orchard easily, some growers choose to widen row spacing and tighten the distance between trees: for example, 18 feet between trees and 24 feet between rows is also commonly employed for tree spacing. Tree density and planting schemes are a direct function of the equipment that will be used to work the orchard.
- Most peach varieties are self-pollinated, so there is no need to arrange trees or blocks of trees to increase fruit set.
- The rule of thumb is that a *minimal* well capacity of 5–10 gallons per minute per acre is required to irrigate a mature orchard nominally. If well capacity does not meet this level, trees will receive less water and the well will probably be running 24 hours a day.
- The average yield of a mature tree is approximately 100 pounds of fruit per tree per year. Some trees are capable of producing several times that amount, but calculate potential revenue based on averages.

- Retail prices can be as much as a tenfold increase over wholesale prices. Some smaller growers net more money than large-scale growers.
- Murphy's law is in full effect: Whatever can go wrong will go wrong. So is O'Leary's corollary: Murphy was an optimist.

Over the years some prospective growers have accused us of being grossly pessimistic about planting a commercial orchard, but our experience is that everyone underestimates how much work is ahead. It is our sincere desire to give you as much advance notice as possible of the risks and challenges that await you as a peach grower. Remember, forewarned is forearmed. Once a decision has been made to plant a peach orchard, each of the subsequent sections should be carefully studied and utilized as much as is practical to ensure continued success.

Chapter 2

. . . .

Site Selection

electing a suitable site for an orchard can provide the greatest chance of economic success. Prospective commercial growers should resist making a site decision based solely on the fact that the land is already owned. If appropriate features are not present, the orchard may be unduly subject to low vigor, soil-borne pathogens, and increased risk of spring frost. It may be more economical to buy new land rather than invest $3,500 per acre and three to four years of time and labor in a venture destined for failure because of poor site selection. Site limitations can be overcome on a limited basis with raised beds. This is often necessary for people who want to plant a few trees and only have 6 inches of soil.

Soils

For an orchard to be successful, the soils must be suitable for peach trees. Trees will be unproductive and short-lived if they are planted in soil types for which they are not well adapted. *Poorly drained sites must be avoided.*

Ideal orchard soil consists of 12 to 18 inches of coarse to medium-textured topsoil, with medium to fine-textured red or brown subsoil. Red and brown-colored subsoils are indicative of well-drained soils, while dull blue, gray, yellow, or white subsoils indicate anaerobic conditions associated with poor drainage.

Drainage can be checked by digging holes 8 inches in diameter and 30 inches deep in several areas of the potential orchard site. Fill the holes

Site
Selection

▲ Standing water indicates poorly drained areas.

◄ Ideal soil profile showing red subsoil.

with water and check the water level after 24 hours. If holes are completely drained, the soil in that area can be considered to have adequate internal drainage. If water is not completely drained after 48 hours, that portion of the site should be avoided. It is best to check for internal drainage during periods of the year associated with high amounts of rainfall so that a worst-case drainage scenario can be detected.

In addition to poor internal drainage, there are other site aspects that may inherently lead to low productivity or short tree life. For instance, land recently cleared of hardwood trees should be avoided due to potential post oak root rot problems. Also, soils with a high pH may have chronic mineral nutrition problems and may be subject to high tree mortality due to cotton root rot.

Each county in Texas has either a detailed county soil survey or a county general soils map. These maps are available at either the county extension office or the county Natural Resource Conservation

Service office. They are now available online at http://websoilsurvey .nrcs.usda.gov/app/HomePage.htm. Study the maps and discuss potential sites with people familiar with area soils. This process can help you eliminate some sites before you even set foot on the property. If soil maps indicate good general soil types, further investigations are warranted.

Raised Beds

If the soil is shallow or poorly drained, raised beds will be in order if planting only a few trees. Commercially it would be best to avoid such sites. A 10 × 10-foot square or a 4 × 8-foot bed built up 12 to 18 inches will work for a single tree. The bed should ideally be filled with the same soil type as surrounds the raised bed. A garden mix can be used if similar soil is not available.

Soil Test

The first step in establishment of a peach orchard is to take a comprehensive soil test. If the soil type is uniform, one composite sample from a minimum of ten different locations in a 10-acre block is desired. However, if obvious differences in soils exist, samples from each

Peach trees do well in raised beds.

distinct area should be taken. This will enable you to detect differences in the surface and subsoils. Take a sample from 1–6 inches and from 6–12 inches down. To do this, dig a hole 6 inches deep with a shovel. Shave a vertical slice of soil 1–6 inches deep from the side of that hole, and place this soil in a clean plastic bucket for your topsoil sample. Continue digging that same hole to 12 inches down, repeat the procedure with soil 6–12 inches deep, and place this in a second bucket for your subsoil sample. Do this at a minimum of eight to ten locations across the potential site. Each bucket of these collective samples should be stirred, and a two-cup portion should be removed for submission to the soils lab for analysis.

The soil analysis will reveal pH, as well as nitrogen, phosphorus, potassium, magnesium, and calcium (N, P, K, Mg, Ca), and some minor elements, all of which are important to the longevity of a peach tree. By taking soil samples pre-plant, any discrepancies can be dealt with before the trees are planted, giving the trees a better start.

The preferable soil pH is 6.0 to 6.5 pre-plant. Where pH readings are lower than this, they can be raised by incorporating appropriate amounts of lime. It is important to incorporate this pre-plant lime as deeply as possible because research from North Carolina has shown that this will increase not only tree growth and yield but also, more important, tree longevity. When the soil pH is above 7.5, iron chlorosis can be expected. In general, nothing can be done to lower the pH in most Texas alkaline soils because of very high levels of calcium bicarbonate. If peaches are planted in soils with these high pHs, expensive annual applications of iron chelates will be necessary. Research and demonstration work continue looking for rootstocks that will overcome this difficulty, and some promising new hybrids are on the horizon.

Many East Texas soils are low to deficient in potassium. The soil level of potassium should be brought

Soil analysis can provide useful information.

up to a high level before planting since this element moves slowly in the soil. If the soil test shows an already high level of potassium, none should be added pre-plant.

Phosphorus is a much more difficult element on which to make recommendations for fruit trees. It is usually difficult to show a tree response with applications of phosphorus at any time other than pre-plant. For this reason low levels of this element should be adjusted in the soil pre-plant, even though the likelihood of seeing a response is limited. Pre-plant fertilizer should be thoroughly incorporated before planting. No nitrogen should be applied pre-plant.

Air Drainage

When the surrounding terrain is not level, the slope and elevation of an orchard site can sometimes mean the difference between having a crop and not having one. Because cold air is denser and heavier than warm air, cold air tends to flow downhill and settle in low-lying areas or against tree lines on the edge of an orchard. On a clear, calm night, there can be a temperature difference of 10°F between the high point

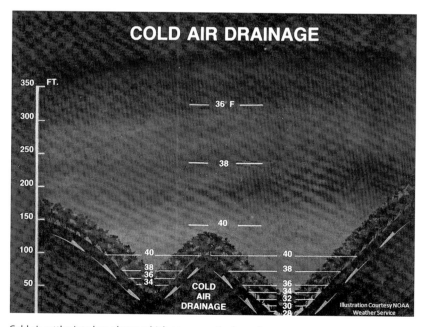

Cold air settles into low places, which can mean the loss of a crop.

and low point of a given site. Trees on the side of a slope can be as much as 5 to 10°F warmer than trees at the bottom of the valley. For this reason, low-lying areas should be avoided unless past experience indicates that spring frost is not a problem.

Questionable sites can be checked by placing shielded minimum temperature thermometers at various locations on calm, cold nights during the fall, winter, or early spring. Do not use windy night readings because the wind continually mixes the air and will not aid in detecting frost pockets. If large differences in temperatures are noted between the highest and lowest areas of the site, the coldest locations should be avoided. If an entire site will be planted, the coldest part of the orchard should be planted in varieties with the highest chilling requirement because they tend to bloom later (chilling is explained in chapter 4).

Probability of Hail Damage

In addition to spring frost, spring or summer hail has the potential to limit peach production significantly. The weather history of an area should be thoroughly investigated before making a decision on a peach orchard site. No site should be planted where the probability of damage from frost or hail is expected in more than two years out of five.

Animal Damage

Many rural areas of Texas have large deer populations. When young fruit trees are planted in these regions, damage from browsing and rutting in the fall can sometimes cause tree loss or loss of a year's growth. While deer repellents are available that work well for short periods of time, heavy deer pressure will require fencing. Certain designs of electric fences are adequate when forage is plentiful, but heavy pressures necessitate a 7- to 8-foot-high deer-proof fence.

On young trees, deer can browse tender foliage heavily, and nutria and porcupines commonly girdle young trees by feeding on trunks and scaffold limbs. Likewise gophers and voles can cause extensive damage by feeding on roots below ground. Buck deer frequently damage tender bark of trunks and limbs with their antlers in the fall. In

Varmint pressure can be a real issue; bird damage just prior to harvest.

bearing trees, raccoons, birds, and squirrels find ripening fruit about a day before you do and can cause extensive loss of marketable crop. Injury and losses from these vertebrate pests must be dealt with on an individual basis.

Chapter 3

Site Preparation

Once a site has been chosen, careful plans should be made to prepare the land before planting to avoid as many potential problems as possible. This preparation should begin the year before planting to allow adequate time for each procedure. Research by the U.S. Department of Agriculture (USDA) in West Virginia, as well as observations here in Texas, indicate that peach trees grow faster when planted on sites that have been in cover crops for one season or several before planting. The research shows that grass roots penetrate the soil profile more uniformly and lead to improved water relations for peach trees planted in such soils.

Strong tree growth on a well-prepared site.

Someone wanting to grow just a few trees can plant trees in a Bermuda grass lawn or pasture and then kill the grass with glyphosate once the grass greens up in the spring. As the grass dies, the roots in the soil decay, which leaves perfect passages for water, air, and nutrients to move into the soil.

Clearing a site of shrubs, cactus, and trees can be accomplished with a bulldozer, and deep chiseling of the site will remove existing roots. Perennial weeds such as Bermuda grass and Johnson grass should be controlled with systemic herbicides such as glyphosate in summer and early fall. Do not use chemical "brush killers" to remove cactus or other hard to control perennial vegetation because these products can leave persistent chemical residues in the soil for years that can kill young trees. Disking between herbicide applications can help control annual grasses and broadleaf weeds for the following years. In some situations, winter cover crops of oats or annual rye grass can inhibit the germination of winter weeds, provide additional organic matter, and help control erosion. Cultivation or herbicide applications will be needed down the tree row in winter to provide a weed-free zone for tree planting, while leaving the vegetation in the row centers will continue to provide erosion control for late winter through early summer.

When replanting an existing orchard site, either annual grasses or sorghum should be planted at least one season before resetting new trees.

Spacing

Many years ago, peaches in Texas were spaced at 30 X 30 feet. These orchards were clean, cultivated, and not irrigated. Air movement between trees was an important component of fungal disease management. Space or land price was seldom a limiting factor, therefore wide spacing was practical.

Today, growers need to plant the trees as close together as possible to realize earlier returns. Dr. J. W. Worthington has shown the optimal tree canopy to be approximately 18 feet in diameter; therefore, this is the recommended in-row spacing, which allows the trees to fill the in-row spacing without overcrowding. The middles between rows should be maintained with a 10-foot sod tractor path down the middle. Hence the optimum peach tree spacing in Texas is 18 feet X 24

Dry land trees were spaced at 30 feet by 30 feet so that the trees could mine large areas of soil.

feet, which gives 100 trees per acre. If drip irrigation is not available or if clean cultivation is to be used, a wider spacing such as 24 X 24 feet will be needed.

Terracing

Peach trees grow better and live longer when planted on terraces (also called beds or berms). While this practice is essential on shallow and poorly drained soils, tree performance will be better even on the best of soils. The terraces can be put up in several ways, with the most common being the use of a bedding plow.

Larger equipment such as road graders can also be used. These will do the job faster and usually more accurately, but may be expensive.

If the spacing is 18 X 24 feet, the terraces should be spaced at 24 feet and the trees set every 18 feet along the top of the terracing. Terrace spacing on rolling land usually requires adjustment to ensure good drainage between terraces. The local Soil Conservation Service office can usually provide assistance in planning the height and specific contour of terraces. Terraces should be 12 to 18 inches higher than row middles. This not only keeps the tree roots out of standing water but

Properly prepared raised beds.

Trees doing well on raised bed contours.

also increases the depth of topsoil under the tree where shallow soil is a problem.

If possible, terrace construction should be done the fall prior to planting so that the soil can settle before trees are set. The sequence of preparation should be: (1) kill perennial weeds, (2) chisel down the row to break up "plow pans," a compacted layer of soil created by frequent prior cultivation, (3) build terraces, and (4) plant winter grass on terraces, then kill it just before planting trees.

Most peach trees planted in commercial orchards are the product of large nursery operations that specialize in fruit tree production. Nurseries are like any other business—the majority of practitioners are honest, reputable producers, but it is best to talk with other growers to find out which vendors have consistently supplied high quality trees that are true to type. Nurseries should be contacted nine to twelve months before the intended planting date to make sure that the desired combinations of scion (fruiting variety) and rootstock are available.

Site
Preparation

Trees are graded and sold either by height or by trunk diameter. While trees are available in a wide range of sizes, growers should resist the temptation to plant the largest trees available. Generally nursery stock 18 to 36 inches high will result in the largest trees after the first year's growth in the orchard. Larger trees often have a larger number of roots cut off as they are dug from the field and suffer greater transplant shock than smaller nursery stock.

Peach trees should be planted as soon as possible after they are received from the nursery. If roots appear dry when the trees are received, soak them in water at least 30 minutes before heeling in or planting. A heeling bed is a trench made in the ground for temporary storage of nursery stock until they can be planted. **Do Not** store trees in water for more than one hour, as this is likely to lead to significant tree loss. Young tree roots can be injured from extended storage in water because oxygen is excluded from this tender tissue, causing death. If they cannot be planted immediately, they should be heeled in.

A heeling bed can be prepared by opening a trench, laying the trees at a 45-degree angle, and covering the roots with soil. The heeling bed should be watered frequently enough to prevent the roots of the trees from drying out. Do not use sawdust or coarse mulch to cover trees because these materials do not hold enough water to keep tree roots moist. Laying trees at a 45-degree angle helps suppress bud break; trees will keep well in this manner but will eventually begin to sprout when the weather warms up. Trees can be stored in a heeling bed for weeks, but the goal is to plant young trees before they break dormancy.

Well-packed bare root trees.

Heeled-in trees will keep for several weeks if the trees cannot be planted as soon as you get them.

Tree Planting

Holes can be dug by any means practical. The most commonly available tools are posthole augers and shovels. Growers in other states are using mechanical tree planters with success. If a large acreage is to be planted, it might be worthwhile investigating one of these implements, since three workers can plant up to three thousand trees per

Set the tree at the same depth as when it grew in the nursery.

day using this method. When using posthole augers, it is important that the soil not be too wet if it contains significant amounts of clay. Augers cause a glazing of the sides of a hole in wet clay or clay loam soils. This glazing increases the resistance of the soil to root penetration at the interface of the soil and the planting hole. The increased resistance reduces the ability of peach tree roots to grow into the surrounding soil. This problem can be avoided by not planting in wet soils or by welding a one-inch protrusion on each side of the auger to score the side of the hole. Hole size is relatively unimportant as long as the hole will accept the root system adequately.

Never place fertilizer in the planting hole. Fertilizer, especially nitrogen, placed in the planting hole with the roots often results in tree death. The young tender roots simply cannot handle direct contact with organic or inorganic sources of nitrogen. Do not apply fertilizer until new growth is 10 inches long; usually April or May.

Planting depth is critical and should be carefully observed. The bud union should not be used as a depth guide, since trees are not all budded at the same height. A peach tree should be planted with its roots at the same depth at which they grew in the nursery. Research from Virginia and Michigan shows that trees are stunted and tree loss is increased if the planting depth is more than 2 inches deeper than nursery depth. Be particularly careful to check depth of planting after a week, as settling often occurs. Depth can be adjusted at this time by carefully pulling the trees up slightly. When filling the hole, soil should be replaced in the same order it was removed. That is, subsoil in the bottom and topsoil on top. If this is not done, an artificial situation is created that can cause problems. Pack the soil firmly around the roots of the young tree and water as soon as possible.

After planting, 30- to 36-inch nursery stock is pruned to a single trunk and headed back to a height of 24 inches tall. All branches are removed, and the lower 18 inches of trunk are wrapped with aluminum foil, felt paper, or any other opaque material. Wrapping young trunks deters shoot growth below desirable scaffold limbs, prevents injury from sunscald and vertebrate pest feeding, and allows for the use of contact herbicides.

Care should be taken to not have any foil around the trunk below the soil surface. This can result in the tree growing into the foil and can create a lesion below ground where pathogens may infect young trees.

Within a few weeks after growth begins in the spring, select the

strongest three to five shoots arising from the top 6 inches on the main stem. They should be evenly spaced along the trunk with at least one directed into the prevailing wind. Remove all other shoots along the trunk or limbs. These few branches will grow vigorously for about four more weeks and then begin to lignify (harden and turn brown) near the trunk. Then is the time to select just three major scaffold limbs. Remove all other upright shoots or scaffold limbs. Continue to let the three major limbs grow, and leave side shoots that are not competing for growth.

Properly planted and wrapped tree.

Strong tree growth showing excellent scaffold limb development.

Weed Control Issues

One of the most critical phases of first-year peach tree care is weed control. Left unchecked, weeds can cause the loss of first year's growth. Most grasses and broadleafed weeds are more aggressive than newly set peach trees in removing both water and nutrients from the soil. Often these small trees can be seen sitting in lush green grass with the telltale red spots of nitrogen deficiency on their leaves.

Whether mechanical or chemical, weed control will greatly increase tree growth during the first year. Mechanical control has the advantage that no chemical toxicity is introduced, and control can be done by unskilled labor when available. Its disadvantages are that frequent cultivation is required for adequate control, and root damage can be extensive if cultivation is too deep. Implements should be adjusted so that they cut no more than 3 inches deep. In this manner, extensive root damage is avoided. Resist the temptation to use weed-

Trees struggling due to weed competition.

eaters around trees for weed control. Even the most careful operator will hit young trees and cause girdling to the trunk. At best, girdled young trees will be stunted, but they commonly become compromised to pathogens or simple physical injury.

As safer chemicals have become available, chemical weed control has become the method of choice for most growers because it is more reliable, does a better job of controlling perennial weeds, and usually does not have to be repeated as often. For specific materials and times of application, refer to the current pest management guide. The best publication currently available is the *Southeastern Peach, Nectarine and Plum Pest Management and Culture Guide*. This guide is updated annually and posted online. It can easily be found by entering the title into a search engine. One note of caution. **Do Not** use Roundup around first-year trees unless the trunk has been wrapped with aluminum foil or some other impervious material. Glyphosate, the active ingredient, is a systemic, nonselective contact herbicide that is relatively safe to the applicator and has little environmental impact. It can be absorbed by all green tissue and cause severe injury or tree death, hence the critical importance of protecting the lower trunk of young trees. The wrap prevents absorption until the trees are old enough for the bark to no longer take up herbicides.

Chapter 4

. . . .

Chilling, Dormancy, Hardiness, and Phenology of Bloom

ature provides internal mechanisms for deciduous trees to prepare for and survive winter and to delay bloom so that there is a
greater chance of successful reproduction. Understanding these
phenomena and managing orchards with this in mind can improve peach tree fruitfulness and tree health, increasing the productive life span of your orchard.

Winter Chilling

Along with site selection, choosing varieties adapted to a given area
of the state is critical for consistent cropping. Choosing varieties with
the appropriate chilling requirement is important in achieving this
goal. As buds are initiated in the summer and continue to develop
in the fall, plant growth inhibitors are produced and translocated to
these buds to inhibit them from growing in the mild weather of autumn. Over the course of the winter, cold temperatures break down
these growth inhibitors, and once the chilling requirement has been
met, the trees can then respond to warm weather and initiate both
floral and vegetative growth.

Fruit varieties commonly have a chilling requirement listed in the
nursery catalog beside their name. This number refers to the number
of hours of winter chilling a variety needs in order to break dormancy, flower, and grow normally in the spring. There are several ways
to measure the accumulation of winter chilling. The old traditional

method is simply to count the number of winter chilling hours at or below 45°F. In more recent years, pomologists have begun counting the chilling hours at the first freeze in the fall. A number of years ago scientists began to investigate the efficiencies of different chilling temperatures in regard to the breaking of dormancy. What they found changed the way we think of dormancy and helps to explain some of the plant response we see in some years.

Effectiveness of temperature ranges on overcoming winter dormancy.

Temperature Range	Chilling Effect
< 33° F	-0-
34°–36° F	+
37°–48° F	++
49°–54° F	+
55°–60° F	-0-
61°–65° F	-
> 65° F	--

What we now know is that temperatures at or near freezing really have no effect at overcoming dormancy. Temperatures slightly above freezing have some effect, but the temperature range between 37° and 48°F is the most efficient at breaking down plant growth inhibitors. Temperatures slightly above that have some effect, but once temperatures reach 55°F, chilling is no longer being accumulated. To some extent, temperatures between 61° and 65°F, and to a greater extent, temperatures higher than 65°F can actually negate some chilling that has previously been accumulated. This phenomenon is much more pronounced during the later part of winter as opposed to during warm weather that may follow the first freeze in the fall. Temperatures in the 80s and 90s in January and early February can dramatically disrupt a tree's ability to break dormancy successfully and can lead to an insufficiency of chilling. It is also worth noting that when we measure temperature effects on dormancy, we measure ambient air temperature. The critical temperature, however, is the temperature inside dormant buds. This means that a 45° clear, sunny winter day is very different from a foggy, rainy 45° day. As growth inhibitors are broken down by winter chilling, they become water soluble, and cold, rainy days have the added benefit of leaching these compounds out of dormant buds. The old adage is that those cold, rainy days are best loved by fruit growers and duck hunters.

Texas and other growing regions across the South can have widely varying chilling zones. The accompanying map shows the average number of winter chilling hours expected at different locations across the state.

Average number of
winter chilling hours
at or below 45°F in
Texas.

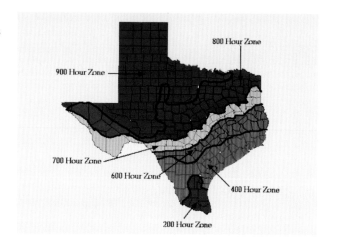

Remember—averages are made up of extremes. A selected location that over the past fifteen years has averaged 800 hours of chilling has in some seasons received as many as 1,600 hours and as few as 600 hours. What are the consequences of choosing a tree with the wrong chilling requirement for a given area? For example, a grower may decide to make sure of getting plenty of chilling and may choose a 500-hour variety for a 700-hour chilling zone. Lower chilling varieties usually bloom earlier than higher chilling varieties. In that case, the 500-hour variety most likely will have its chilling requirement fulfilled very early in winter. In that case the tree will respond to the first warm spell and begin to bloom. Varieties with too low a chilling requirement for a given area typically bloom far too early and are at a much higher risk of being frozen out.

If a grower selects a very high chilling variety in order to reduce the spring frost risk there could also be negative effects. When a tree receives insufficient chilling, dormancy is not properly broken, trees are very slow to leaf out, and they have little or no crop. With insufficient chilling, flowers are poorly developed and are not functional. Although flower buds are initiated the previous summer, buds finish developing or differentiating only after they have received their full amount of winter chilling in early spring. Like the floral buds, vegetative buds also have a chilling requirement. When trees do not receive sufficient chilling, shoots are extremely slow in forcing. This can result in trees with little or no foliage in late spring. This is incredibly stressful for the trees because without foliage, trees cannot photosynthesize food for growth and are subject to sunscald from the hot summer sun. With little crop, trees amply supplied with water, nutrients, and minimal weed competition commonly recover by midsummer.

Insufficient chilling is a cumulative stress. With trees drastically out of place relative to their chilling requirement, insufficient chilling will eventually result in tree death. Commercial growers do have a growth regulator named Dormex that can help flowers and vegetative buds finish developing and force normally when there is a small insufficiency in chilling. Dormex is a restricted use product and is very unstable unless stored in refrigeration. Contact your local AgriLife Extension Service agent or specialist if you would like more information on the use of this product.

Site Selection Can Mitigate Frost and Freeze Injury

Proper orchard site selection is the single greatest factor for mitigating the risk of spring frost. Cold air is denser than warm air and consequently moves to low-lying areas such as creek and river bottoms on still mornings. Because peach trees are relatively early bloomers, such deep bottomland sites should be reserved for late-emerging crops such as pecans or persimmons. Elevated, sloped fields that facilitate the movement of cold air away from trees on a still, frosty morning are preferred for peach orchards. In some cases, native tree lines may

With insufficient chilling, flowers do not differentiate fully and set little or no fruit.

With dormancy not properly broken, trees are very slow to produce foliage in the spring.

hinder air movement away from the orchard site, and these trees may need to be totally or partially cleared in order to do away with this "cold pocket." It is important to note that the temperature on the surface of fruit can be lower than ambient air temperature when frost occurs. Especially under low relative humidity, flower and fruit tissues radiate heat, which is dissipated into the atmosphere, and plant tissue can actually become colder than the ambient air temperature. Fog, high humidity, or cloud cover can help mitigate this effect, but on dry, still, cold mornings, frost can occur with ambient air temperatures as high as 35°F.

Winter Hardiness

Because deciduous fruit trees evolved in temperate climates, they respond to environmental events in order to prepare for winter. The process of a plant becoming hardy starts in midsummer after the summer solstice, when trees respond to the shortening photoperiod (day length). As days become shorter, trees cease vegetative growth and form terminal buds on the apex of seasonal growth. As late summer progresses into fall, annual growth becomes more woody and takes on a dark mahogany color, and the presence of initiated flower buds becomes more evident at the base of each leaf. As days continue to shorten and temperatures cool, leaves begin to senesce, and an abscission zone begins to form where leaves are attached to stems. The first fall frost is a strong signal that triggers a hardening of plant tissues. As temperatures get colder and colder, plant hardiness increases until it reaches its maximum hardiness.

There is tremendous variability among peach varieties in their maximum hardiness levels; indeed, we grow some of the same varieties in Texas as are grown along the Great Lakes. The question then becomes: Why do we have winter hardiness problems in Texas with the same varieties that do not suffer winter injury in Michigan or New York? The answer lies in the fact that hardiness can be gained and lost over the course of the winter, depending on temperatures preceding a freeze event.

The rule of thumb is that plants gain hardiness very slowly but can lose it very rapidly. Often our winters are interrupted by periods of warm to hot conditions in January or February. When this happens, peach trees lose much of the hardiness they have gained. If temperatures slowly return from cool to cold, plants reacclimate and can be fully hardy again. Unfortunately, the interruption of hot winter temperatures is often halted by a big northern cold front causing temperatures to plummet once again. When this happens, peach trees can suffer mild to severe winter injury. Winter injury commonly occurs on the south or southwestern side of trunks or on the upper regions of major scaffold limbs. This is because on a cool or cold clear day, radiant energy from the sun strikes the bark of trunks and scaffolds and warms the tissue beneath the bark. We feel the same phenomenon when in a car on a cold clear day. Even though the ambient air temperature or the temperature of the windshield may be cold, we feel the warmth of the sun striking us through the windshield. When this happens to trunks and limbs, the tissue that becomes warmed suffers a localized loss of hardiness. When the sun goes down at dusk and the air grows cold, those compromised tissues are at greater risk of winter injury because of nighttime freezing temperatures.

The next question is: What can we do to mitigate freeze injury? The obvious answer is that we need to take the best care of our trees

Winter injury to trunks usually occurs on the south or southwestern side of the tree.

Sometimes mistakenly called sunscald, winter injury usually occurs on the upper surface of scaffold limbs. As shown in this photo, with proper tree health, callus tissue can heal such injury and the limb can recover.

that we can. As discussed in the pruning and training section of this book, growers need to think of carbohydrates, the products of photosynthesis, as the energy currency of their trees. It is generated through photosynthesis, and the healthier a tree is, the more efficient the tree is at generating carbohydrates. Trees expend this currency for initial growth in the spring, but the main expenditure of this energy is through fruit production. If a tree overbears, it becomes overdrawn at the bank and bad things happen. Maintaining a healthy canopy as late in the fall as possible maximizes postharvest photosynthesis, which can lead to what plant physiologists call carbohydrate loading. This means that carbohydrates become stored in roots, trunks, limbs, and buds within a tree and that the tree is in good condition going into the winter.

Carbohydrate loading is also extremely important in minimizing winter injury because carbohydrates are also the anti-freeze mechanism of the tree. Carbohydrates are complex sugars. When one places a glass of water in the freezer at 32°F, ice crystals begin to form. If a teaspoon of sugar is added to the water and dissolved, the freezing point is suppressed. If one continues to add sugar until no more sugar can be dissolved, we consider this a saturated solution, and the freez-

Keeping a tree's canopy healthy as late in the fall as possible is important in optimizing tree winter hardiness.

ing point is as low as possible. The same principle applies to plants. If we do not overcrop our trees, but do keep weeds under control and make sure our trees have ample water and nutrients, the canopy will remain healthy well through the fall.

This optimizes carbohydrate storage and winter hardiness. Many times, especially in young, nonbearing orchards, leaves may remain healthy on these trees even with temperatures dipping into the mid to low 20s. This again is because the leaves are full of carbohydrates and do not freeze at 32°F. We can do nothing to affect the weather to which our trees are exposed, but we should prepare them as best we can for the unpredictable winters we have in Texas.

Variety Placement

Growers typically choose a sequence of fruit varieties that ripen over a season and have chilling requirements that approximate the average winter chilling a location receives. As already noted, not all varieties have the same chilling requirement. Because lower chilling varieties usually bloom earlier than higher chilling varieties, placing low chilling varieties at the highest point of a site may help prevent damage by freeze or frost.

Orchard Floor Management

The condition of orchard floors impacts their ability to receive, store, and deliver heat to trees in an orchard on a frosty morning. Orchard floors that have dead weed cover or that have mulch under the trees are insulated from heat delivered on clear days during the spring. In other words, the heat is reflected by the mulch or grass cover and is not absorbed by the soil. This can keep orchard floors cool and delay bud break, thereby increasing the probability of avoiding frost problems. Sandy orchard sites usually warm earlier in the year and typically bloom a little earlier than clay or sandy clay orchard sites.

Once peach trees begin to respond to warming spring conditions the dynamics change. After bloom begins, a weedy or mulched floor will absorb less heat during the day and have less heat to release on a cold morning. Weedy and mulched orchards will be colder on a frosty morning than an orchard with bare soil. Clean, cultivated orchards, especially sandy ones, are the first to respond to warm spring temperatures and bloom first, ahead of orchards on soils with higher clay content and mulched or weedy orchard floors. While there may be other reasons to avoid cultivation, especially deep cultivation, there is no denying that once bloom has begun or has passed, a bare orchard floor will be warmer on a frosty morning than a mulched or weedy floor. A bare, clean, well-packed herbicide strip under the tree row is perhaps the best at absorbing and releasing heat.

We suggest that growers utilize floor attributes both to deflect and absorb heat. We suggest planting and maintaining a winter cover crop in row centers until bloom begins, then employing tight, frequent mowing of row centers to make the orchard floor more capable of absorbing and slowly releasing heat. We strongly recommend having a bare, clean herbicide strip as opposed to cultivation under the trees.

The Use of Water to Prevent Frost Damage

Water can be used in two ways to prevent crop loss due to spring frost or freeze. For full understanding of these techniques, we need to explain briefly the physics of water evaporation and freezing.

A bare strip under trees at bloom will give them the greatest amount of potential protection from spring frost.

Evaporation is an endothermic reaction. That is, when water evaporates, it receives heat, or has a cooling effect on the tissue from which evaporation takes place. We all know how refreshing it is to be sprayed with a garden hose on a hot day; the same effect takes place when water is delivered to dormant trees via overhead irrigation. Water applied as a mist or timed coarse spray from overhead irrigation devices can slow bud advancement, which can delay bloom and give these trees an advantage in avoiding injury from spring frost. Seven- to ten-day delays in bloom have been documented, and this technique may be a management strategy for growers who have ample water and choose to invest in a separate overhead irrigation system in addition to drip or micro-sprinklers that are used to deliver water to trees during the growing season. However, there are no guarantees in this business; there have been cases when bloom has been delayed for two weeks, only to have the crop freeze out in week three.

Water has also been used in other perennial crops to insulate developing buds or fruit from late season freeze or frost. In contrast to the cooling effect of evaporation, when water freezes, there is an exothermic reaction in which the freezing water releases heat and insulates tissue by coating it with ice. Ice freezes at 32°F, and if water can be used to maintain that temperature, tissue injury can be avoided.

Overhead sprinklers can help protect open blooms by encasing them in ice on a cold morning. Ice remains at 32°F and can insulate from colder temperatures.

The critical temperature for lethal freezing of buds and young fruit differs by variety and growth stage. Generally, the lethal temperature for 50 percent of flowers in full bloom is thought to be about 28°F. Buds less developed are hardier, but once fruit is set, many fruit can be killed at 30°F. If the application of water keeps the flower or fruit tissue at or near 32°F there may be minimal crop damage. Growers should be aware that the application of water on a frosty night must begin well before temperatures reach freezing in order to provide protection. This technique may also have the drawback of causing tree injury. Heavy ice loads may break scaffold and sub-scaffold limbs.

Weedy orchard floors are much colder during a frosty morning than are cultivated orchards. A clean, weed-free orchard floor receives and stores the maximum amount of solar radiation and releases it slowly, giving these sites the best possible means of avoiding freeze damage.

Taking Advantage of Inversion Layers

During the day the sun's radiant heat warms the soil. As soil absorbs heat, it becomes warmer than the air with which it is in contact, that heat is transferred to the air, and the air begins to rise. At day's end the heat is predictably spread throughout the atmosphere with the warmest air at the soil surface and temperatures decreasing with height in the atmosphere. After sunset the ground begins to lose heat through radiation and becomes colder than the air with which it is in contact. The release of heat from the soil is usually limited to a few hundred feet above the ground. The zone between rising warm air and the normally cold atmosphere is called an inversion. Since warm air is lighter than cold air, the warmest air is at the top of the inversion. A number of atmospheric factors determine the height of an inversion, but when it is relatively low, it can be employed, through the mixing of air, to mitigate spring frost.

Orchard Heaters

In the past, when energy was cheap, orchard heaters were commonly used to increase the amount of heat within an inversion layer. If the inversion was low enough, and enough heat was produced, the tem-

Small fires placed around the orchard floor can help warm trees and mitigate frost injury.

perature of the orchard could be raised enough to be economically beneficial. Return stack heaters fueled by diesel oil have been used by citrus and tropical crop growers as well as peach growers. Others have used coal, small fires, and burning hay bales to provide heat. Remember, though, it is the heat and not the smoke that provides the protection. One heater or small fire per twenty trees is standard.

Fires must be small for heat to be retained within the inversion layer. More is not necessarily better, because large fires can produce such a large plume of heat that it can puncture the inversion layer and be entirely lost. Environmental concerns and fuel prices have brought a decline in the frequency of growers using heat to protect orchards in the last two decades.

Mixing Air

Mixing the air within the inversion layer in order to move the warm air at the top of an inversion layer back to the orchard floor can also provide significant warming effects. Fixed-in-place wind machines with rotating blades have been employed for this purpose, as have helicopters. Both techniques can be expensive, but remember, in many cases raising the temperature only a couple of degrees can mean the

Phenology of Bloom

Peach tree buds begin to form on new shoot growth about the end of May. These buds are strictly vegetative and are useful in asexual propagation (budding) of new peach trees. By late July or early August floral buds begin to be noticeable at the base of leaves on the current season's growth.

As late summer and autumn progress, the annual growth turns from green to a mahogany brown. Leaves abscise and the tree prepares for winter. After the chilling requirement has been broken and temperatures warm, flower buds begin to swell.

The stages of bloom go from first swell to popcorn

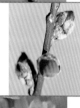

to pink bud, where the first color of the petals becomes visible.

Petals continue to enlarge and protract to the stage of bloom known as balloon stage, right before the flower fully opens. Anthesis (pollen shed) typically begins at this point, and many peach flowers are actually pollinated before the flower opens. The next photos show flowers in full bloom. Some varieties, such as 'Loring' and 'Redskin,' have flowers with large petals and have what are known as "showy" flowers.

Others, such as 'Sentinel' and 'Harvester,' have flowers with smaller petals, which are known as "nonshowy" flowers.

After pollination the flower begins to turn into a small fruit, and the petals drop as the tree enters into the growth stage known as petal fall.

At this point cell division results in a rapid stage of growth for young peaches, and they gain size very quickly. As the size begins to exceed that of the flower's ovary, the corolla or calyx begins to crack, giving way to the new fruit. This growth stage is known as shuck split and is important in timing sprays to control plum curculio and peach scab.

Wind machines can mix warm air at the top of an inversion layer, warming temperatures at tree height.

difference between having a crop and freezing out. Even though expensive, mixing air can pay large dividends if the temperature of an orchard can be kept above the critical 30°F mark. Helicopters must continue to mix air and wind machines must continue to operate until ambient air temperatures rise above freezing, sometimes hours after daybreak.

Last but not least, for the homeowner with only a couple of trees, placing tarps or blankets over the tree can help trap escaping warm air and keep the tree from being injured in mild frost events. Additionally, a small incandescent light can provide a small constant amount of heat that will be caught in the canopy of the tree. With frost events there is usually no wind, so these tarps do not need to be secured during the night. Care needs to be taken, however, because when the temperature rises the following morning, the wind will begin to blow and if not taken off in time, the tarp itself can do considerable damage to a tree with blooms or small fruit.

Chapter 5

· ·

Cultivar and Rootstock Recommendations for Texas Peach Growers

Commercial peach production has been important in Texas since the 1890s. Peach cultivars are generally self-fruitful and generally do not need other varieties as pollenizers (sources of pollen). To be successful choices for a given location, varieties should produce a reliable crop of fruit that is firm and has commercially acceptable size, an attractive blush, and good eating quality. Growers typically choose a number of varieties that ripen in sequence so that fruit is available for an extended period over the course of the season. The varieties that are suggested are listed by chilling category—high, medium, and low chilling—and adapted to various parts of the state. A map of approximate average winter chilling accumulation appears in chapter 4 of this book. Although varieties are categorized in high, medium, and low chilling groups, growers should pay specific attention to the listed chilling requirement to determine adaptability to a given area. Varieties are described in ripening order relative to the age-old standard 'Elberta.' Although it has been decades since that variety has been recommended for commercial production, it remains an important standard in knowing when varieties in high and medium chilling zones ripen. For lower chilling cultivar comparisons, 'Flordaking' is used as the standard.

Varieties are also described as to their stone freeness. Typically, very early season varieties are clingstones, midseason peaches are described as semi-cling or semi-free, and later season varieties, which have a definitive cavity between fruit flesh and the pit, as freestone fruit. Some varieties are physiologically clingstones, even though they may ripen late in the season.

Selected High Chilling Variety Descriptions

'**Regal**' is a relatively new variety that ripens very early in the season. The product of a cross between 'Surecrop' and 'Harvester,' 'Regal' was released by the Louisiana Agricultural Experiment Station in 1992. It is very resistant to bacterial leaf spot and has greatly improved size and quality over 'Springold,' the old standard for that ripening slot.

'**Goldprince**' was first released in 1989 from the Southeastern Fruit and Nut Research Laboratory in Byron, Georgia. It is moderately resistant to bacterial spot and is the replacement for 'June Gold,' which may have an excessive amount of split-pit fruit.

'**Sentinel**' is a variety with both positive and negative attributes. The ridge or suture on the side of the peach can be very prominent, especially in low chilling years, making this variety hard to ship without excessive fruit bruising. This variety also has a high degree of pubes-

Cultivar and Rootstock Recommendations

High chilling varieties (700- to 1,000-hour zones).

Variety	Chilling Requirement	Stone Freeness	Days Before Elberta
'Flavorich'	700	Cling	64
'Regal'	700	Semi-Cling	54
'Goldprince'	650	Cling	46
'Surecrop'	1000	Semi-Free	42
'Juneprince'	650	Semi-Free	35
'Sentinel'	850	Semi-Free	34
'Gala'	750	Semi-Free	34
'Harvester'	750	Free	26
'Ranger'	1000	Free	24
'Fireprince'	850	Free	20
'Cary Mac'	750	Free	20
'Topaz'	850	Free	18
'Majestic'	850	Free	16
'Redglobe'	850	Free	13
'Cresthaven'	850	Free	3
'Dixiland'	750	Free	3
'Redskin'	750	Free	2
			Days After Elberta
'Flameprince'	850	Free	14
'Parade'	850	Free	30
'Fairtime'	750	Free	35

'Regal'

'Surecrop'

'Sentinel'

'Harvester'

'Majestic'

'Redglobe'

'Redskin'

'Flordaking'

'Texstar'

'LaFeliciana'

cence or fuzz. That said, 'Sentinel' is a reliable producer of high quality fruit. It may be best for roadside or pick-your-own operations.

'GaLa' is a 1994 joint release from the Louisiana Agricultural Experiment Station and USDA's Agricultural Research Service. It ripens with 'Sentinel' and was released to be a firmer, less pubescent replacement for it. It is not consistently productive in some locations.

'Harvester' is a highly recommended standard midseason variety grown across much of the state. The fruit is round, firm, and good quality, and the tree is vigorous and very productive. This is the first freestone of the season in this chilling regime.

'Dixiland' is another standard variety that ripens relatively late in the season. It has a reputation of being a consistently productive, high quality variety that ripens at a time when there are several other varieties also maturing.

Selected Medium Chilling Variety Descriptions

'Flordaking' is an important early season variety for moderate chill zones. It is a consistent producer of large, somewhat soft fruit most adapted to local sales. It is moderately resistant to bacterial spot and, like other varieties, it can have a prominent suture and tip, especially in years with marginal chilling.

'Juneprince' is a moderate yielding variety that produces firm, attractive fruit. The fruit is average to large in size, firm, and ripens with an attractive blush. The trees are moderately susceptible to bacterial leaf spot.

Medium chilling varieties (450- to 650-hour zones).

Variety	Chilling Requirement	Stone Freeness	Days Before Elberta
'Flordacrest'	425	Semi-Cling	55
'Flordaking'	450	Cling	51
'Goldprince'	650	Cling	46
'TexKing'	450	Cling	42
'Juneprince'	650	Semi-Free	35
'Texstar'	450	Semi-Free	32
'Southern Pearl'	650	Free	28
'TexRoyal'	600	Free	25
'Suwanee'	650	Free	22
'TexPrince'	550	Free	20
'La Feliciana'	600	Free	18

'**Southern Pearl**' is a white-fleshed variety. Because of the tendency of white-fleshed peaches to be soft and easily bruised, the old adage is that "there are two days that everyone wants a white-fleshed peach—yesterday and tomorrow." 'Southern Pearl' breaks that stereotype and is growing in popularity in medium chill zones. It has good firmness, productivity, and fruit quality.

'**TexPrince**' is a midseason, moderately productive variety that ripens shortly after 'TexRoyal.' The fruit is firm and of good quality and ripens with a high percentage of blush on the skin. Trees are vigorous and moderately resistant to bacterial spot.

'**La Feliciana**' is a release from the Louisiana State University breeding program at Calhoun and has long been a standard for moderate chilling zones. Relatively late to ripen, it extends the season for growers in chilling zones of 500–600 hours. Fruit is moderately large and somewhat soft. It is good for local markets but not a dependable shipper.

Selected Low Chilling Variety Descriptions

'**Gulfking**', like many of the lowest chilling peach cultivars, came out of the University of Florida's peach breeding program. 'Gulfking' ripens early and has a "nonmelting" flesh that helps maintain firmness after harvest. Fruit is attractive and has good quality if fully tree ripened.

Low chilling varieties (150- to 400-hour zones).

Variety	Chilling Requirement	Stone Freeness	Days Before 'Flordaking'
'Gulfking'	350	Cling	6
'Flordacrest'	375	Cling	+4
'Gulfprince'	400	Semi-Free	+25
			Days After 'Flordaprince'
'FlordaPrince'	150	Cling	-0-
'TropicPrince'	150	Cling	+7
'TropicBeauty'	150	Semi-Free	+14
'TropicSweet'	150	Free	+22

'**Gulfprince**' is at the higher end of chilling for this grouping; fruit ripens in early June in most 400-hour locations. The fruit is of moderate size with a dark yellow ground color and attractive red blush. Another nonmelting flesh variety, it has good firmness but is best for local sales.

'**TropicPrince**' is a very low chilling 1994 Texas A&M release. Fruit has good size, color, firmness, and quality. Adapted to subtropical conditions, this variety is for those in deep South Texas or on the lower Texas coast.

It is also important to understand that some varieties are far more flexible in their ability to set a crop under high and low chilling conditions.

Suggestions for Homeowners

Homeowners often have room for or are interested in having only a single peach tree in their yard for personal needs. Choosing a single best variety to recommend for each of the chilling zones is a difficult task, but for the higher chilling zones of Texas, 'Harvester' has been shown to be the highest quality, most consistently productive variety over time. 'Harvester' is a midseason freestone that is excellent for fresh eating as well as for freezing or canning.

For medium chill regions 'LaFeliciana' would appear to be the best choice. 'LaFeliciana' is a freestone variety that ripens in midsummer with good quality and is easy to process for enjoying later.

For low chilling zones, 'Flordaprince' is perhaps the variety with the greatest track record of consistent production. It has good quality and is the standard by which new low chilling breeding selections are measured. ■

Consult your local county extension agent, county horticulturist, or extension specialist to discuss the strengths and weaknesses of specific varieties for your area and your marketing strategies.

Peach Rootstocks

Rootstocks are used to overcome a number of challenges that soils, pests, and pathogens pose to the health and longevity of peach trees. Finding the right variety on the right rootstock can be challenging, but every effort should be made to search out nurseries that propagate exactly what you want. In some cases, this may mean significant advance planning, a year and a half before you intend to plant. Rootstocks adapted to the larger production regions of the southeastern United States are not a good fit for most Texas locations. Many Tennessee nurseries that used to supply a majority of peach trees to Texas growers are now propagating trees only on Guardian rootstock. Consequently, more peach trees are now being bought from nurseries in Texas and California.

Nemaguard

Once the only peach rootstock recommended in Texas, Nemaguard is becoming hard to find except with very low chilling varieties. It is valuable because of its tremendous resistance to root knot nematode. It is not especially tolerant of clay soils and can become quite chlorotic on high pH soils because of its inability to take up iron in those settings. Nemaguard can also be somewhat winter tender. That said, for growers on sandy sites with a history or probability of root knot nematodes, it remains the single best rootstock for overcoming this important soil-borne pest.

Lovell (and Halford)

Once popular canning cling varieties grown widely in California, these two are now far more commonly used as rootstocks. While neither provides any protection against nematodes, both are far more adapted to heavy soils than is Nemaguard. These rootstocks are adept at taking up iron on high pH soils and are preferred on clay sites where nematodes are not an issue.

Guardian

Developed and released by the USDA and Clemson University, this rootstock was specifically chosen for its ability to overcome a condition in the southeastern United States known as peach tree short-life. Its tolerance to several species of nematodes gives it the ability to maintain winter hardiness where others have failed. Unfortunately, it appears that Guardian succumbs to the race of southern root knot nematode found in Texas. Under limited trials and observations, this rootstock has performed poorly in both East Texas and the Hill Country.

Interspecific Hybrids

A number of rootstocks have been released that are the product of a peach almond cross. These rootstocks are apparently tolerant of root-knot nematodes but are specifically utilized to overcome micronutrient deficiencies on high pH soils. Unfortunately, these rootstocks are very sensitive to wet conditions; they are relatively winter tender and therefore quite susceptible to bacterial canker. While trees perform well in early years, they tend to be rather short-lived and rapidly lose productivity.

Myrobalan Plum (Prunus cerasifera)

This stock is sometimes used for peaches but is more commonly used for plums. It offers some resistance to poorly drained soils and post oak root rot. It is very shallow rooted and extremely prone to suckering. It is not recommended for commercial orchards but may have a place in overcoming soil restrictions in backyard plantings.

Mexican Plum (Prunus mexicana)

This species of wild plum is unique in that it produces a single-trunk tree without suckering, as most wild plums do. It has some tolerance to cotton root rot, so it may be a possibility where this fungus is a problem. Mexican plum trees would have to be budded to your chosen variety.

Peach Tree Pruning and Training

each trees require annual dormant pruning in order to remain healthy and productive. Before we explore *how* we prune trees, it is important first to examine the growth cycle of these trees and understand *why* we prune peach trees. Peach fruit are borne only on the previous season's growth. So, for all practical purposes, each year we are growing two crops: the fruit that are being produced that season, and the wood that will be responsible for the next year's crop. It is important that as growers, we learn to strike a balance with our trees between fruit production and vegetative vigor. Too much of either is undesirable and leads to poor tree health and a lack of productivity.

Optimizing Light Interception

One of the goals of proper pruning and training of peach trees is to optimize light interception to desirable areas of the tree. Peach flowers are initiated on green shoot growth in the middle of the summer. At that time vegetative and potentially flower buds arise at the axil of leaves that become the new growth points and flower crop for the next season. If a shoot is well supplied with water and nutrients and is exposed to sufficient sunlight, three buds typically form at the base of each leaf: a central vegetative bud and a flower bud on either side of the vegetative bud.

These buds overwinter and, if exposed to sufficient winter chilling, grow the following season into flowers or vegetative shoots. Because

▲ Typical bud arrangement on previous season's wood at bud swell. A flower bud normally is formed on each side of a vegetative shoot although solitary flowers or strictly vegetative buds can also be found.

▶ A typical current season's shoot. Growers strive to produce a canopy of shoots 12–18 inches long each season. Note that the apical tip is still actively growing in midsummer.

of this need for adequate sunlight exposure, one of the major goals of annual dormant pruning is to shape the tree to minimize shade into the fruiting zone and to ensure that adequate sunlight strikes the canopy to optimize photosynthesis. Shaded portions of the tree will die or have very low vigor, so shade must be avoided. Fruit produced in a highly shaded canopy will have poor color and fruit quality. Photosynthesis drives every function of the tree and is referenced numerous times throughout this book.

Imparting Vigor

One of the most important concepts to understand about pruning is how it affects the vigor of trees. Dormant pruning is an invigorating action. The more severely a tree is pruned, the greater the vigor of the return growth the following spring. This response is called compensatory growth. The converse is also true; pruning during the growing season is a dwarfing action. There may be times when this technique can be most useful, and summer pruning is discussed later in this chapter.

As mentioned earlier, each year we are charged with balancing fruit production and vegetative growth on bearing trees. On average, we strive to produce annual shoot growth from 12 to 18 inches in length each year. In addition to stimulating vigor with winter pruning, adequate vegetative growth is accomplished by ensuring that the tree is supplied with adequate water and nutrients. Excessive crop load will retard vegetative growth. Our goal in growing healthy, productive peach trees is to have a tree grow very vigorously early in the season in order to maximize photosynthesis early in the season. We want our vegetative canopy present early to manufacture the photosynthates necessary for optimal fruit growth.

Crop Control

A mature peach tree produces many more flowers than are needed each year for a full crop.

In addition to imparting vigor and reducing shade, pruning serves as the first line of defense against overcropping. During the process of pruning a mature tree, larger corrective cuts are made initially, and the final stages of the pruning process are to thin out annual fruiting wood. Some growers choose to leave considerably more fruiting wood

Mature peach trees can bear in excess of 10,000 blooms per tree each year.

than others as a form of "fruiting insurance" against spring frost. This strategy may pay off in some seasons, but in years with a heavy fruit set, the job of fruit thinning becomes considerably more formidable, time-consuming, and expensive.

Air Circulation

Last, but certainly not least, we prune trees to open them up to allow for quick drying after a rain event. Brown rot, our most serious fungal pathogen that infects fruit, is driven by temperature and canopy wetness. In other words, it may not matter how much it rains—what matters is how long the canopy remains wet. If a tree is well pruned to accentuate air circulation, the canopy can dry quickly and minimize potential disease pressure. While this approach does not usually stand alone as a way to prevent brown rot, it is the one cultural practice we can call upon to minimize chemical inputs or help in brown rot control in high rainfall seasons.

When Do We Prune?

The question of when to prune can best be answered with another question: How many trees do you have? This response is intended to mean that we prune peach trees as late in the dormant season as time allows. Although pruning a young orchard may go very quickly, mature trees may require as much as 20 to 30 minutes per tree. With an average of one hundred trees per acre, it would take 40 to 50 hours per acre to accomplish the task. Homeowners and small-scale commercial growers can easily finish pruning in the spring and can pick and choose the weather in which they want to work. With larger commercial operations, demand for substantial additional labor arises, and there may well be pruning done under less than ideal environmental circumstances. Pruning is truly a combination of both the science and the art of pomology. Many liken it to sculpture in that we need to culture in our mind's eye an image of what a tree should look like, and strive to make the best of what each tree has to offer. Few trees grow exactly as we would want them.

The ideal time to prune a peach tree is during bloom. When trees are pruned during bloom and before leaves emerge, the pruning is still an invigorating action. This timing also lets one see exactly what the production potential is because the blooms are clearly visible. A common mistake among new growers is failing to prune a mature tree severely enough. Pruning during bloom often gives us the courage to prune correctly because we can actually see how many blooms remain on the tree. Dormant pruning can take place any time in the winter once trees go completely dormant. This usually happens after the first hard frost or freeze in late autumn. However, there are at least two reasons to wait until late dormancy to prune your trees.

Bloom Delay

Once dormancy is broken, trees begin to initiate bud growth (both floral and vegetative) in response to warm weather. Trees normally have a progression of bloom from the apical tip of a shoot to the basal end. When a shoot is pruned and the apical blooms are removed, the basal flowers will bloom earlier than they would have if left unpruned. A pruned tree may bloom as much as seven or eight days ahead of an unpruned tree. While this may sound like a small length of time, a week of bloom delay may mean the difference in crop loss due to frost or having a tree escape damage. Lower chilling varieties will have dormancy broken earlier in the winter than higher chilling varieties. When this happens, they begin to respond to warm spells and accumulate heat that promotes growth. This functionally means that lower chilling varieties normally bloom ahead of higher chilling varieties. For that reason, growers not only plant their lower chilling varieties on the warmest location in the orchard; they normally also prune the lower chilling varieties last in order to delay bloom as long as possible. Care should be taken, however, to make sure that dormant pruning is finished by the end of bloom. While some pruning may be done after leaf tissue is present, it is no longer invigorating and may indeed slow vegetative growth.

Avoiding Tree Infection

One of the most debilitating pathogens of peach trees is bacterial canker. Specifics of this disease and other control strategies are discussed in chapter 10, but the time of pruning has a tremendous impact on

the probability that a tree will suffer a sustained infection from this
bacterial pathogen. Pruning wounds should be considered as targets,
and when bacterial inoculum encounters an open pruning wound, infection commonly occurs. If infection takes place in late fall or early
winter, the tree is not in an active stage of growth and has little internal defense against this infection becoming established. Once a tree
is infected, the pathogen moves systemically in the tree and there are
no chemical controls that can arrest the growth. When we wait until
spring to prune, if the same infection event takes place, the tree is
much closer to an active stage of growth and is in a much stronger
position to ward off systemic infection. It is also worth mentioning
that if trees are infected in an orchard, it is wise to sterilize pruning
equipment with either denatured ethanol or a mild Clorox solution
between trees and certainly after pruning through a portion of the
tree believed to be infected.

Summer Pruning

Summer pruning is used to manage the interception of sunlight by
the canopy. It commonly involves removing root-sprouts or suckers
as well as removing some of the highly vigorous, upright vegetative
growth that shades desirable areas of the canopy. Remember that unlike dormant pruning, summer pruning is a dwarfing action.

Growers should consider photosynthesis as a way to generate "currency" in a tree's bank account, and carbohydrates—the product of
photosynthesis—as the money in that bank account. Carbohydrates
are the energy units of the tree. They are produced by photosynthesis
and spent on shoot growth and fruit production. When green tissue
that has been grown by stored carbohydrates is removed with summer pruning, it is a net photosynthetic loss or results in a withdrawal
from the bank. Excessive removal of green tissue, or overcropping a
tree, will leave it overdrawn at the bank and with no way of generating more funds. Carbohydrates have other important functions in
perennial plants, including frost and freeze protection, as discussed in
chapter 4.

Summer pruning is most effective when conducted early in the
summer and may need to be done in several passes through the orchard. Excessive removal of interior green tissue can severely dwarf

With vigorous growth, peach trees may need some excessively vegetative tissue removed from the center of the tree to optimize light exposure to the desirable bearing canopy. Care should be taken to not overdo it and stunt or sunburn the trees.

the tree and can also expose previously shaded scaffold limbs to excessive solar radiation and result in sunscald of the limbs. Summer pruning opens up the tree canopy and can decrease bacterial leaf spot disease pressure in exceptionally wet springs and early summers. Some observers believe that summer pruning can increase bacterial canker infections, but this is generally the case when summer pruning is done in late summer or early fall. Summer pruning should be concluded by mid-July because pruning after that will not provide additional light to the remaining canopy prior to the conclusion of bud differentiation. Removal of root-sprouts should be done as early in the growing season as is practical.

Pruning in the fall is not advised. We strive to maintain an active, healthy canopy as late in the fall as possible, and fall pruning can result in severe bacterial canker infection.

Pruning Basics

Although a number of different training systems can be used to train peach trees, the "open center" or vase system is the most common in

Root-sprouts, or growth coming from below the graft union, should be removed to keep growth in the desirable portion of the tree.

the southern United States and is the system presented in this book. Unlike pear and pecan trees, which have a very strong upright stature, peach trees have more of a "willowy" growth habit that lends itself to open center training. The goal of this training system is to establish a single trunk approximately 18 inches high and then establish three or perhaps four scaffold limbs that support the bearing structures of the tree. Let us start at the beginning.

Pruning at Planting

When a tree is grown and dug in the nursery, a certain percentage of the root system is lost in the digging process. At planting, we prune only roots that appear to be either damaged or potentially diseased. Roots are an important carbohydrate storage site and should be left intact as far as possible in order for young trees to get off to a strong, healthy start. We do, however, prune the above ground portion to put the top growth and root system back into balance. To establish a head or crotch of the tree with primary scaffold limbs at approximately 18 inches above the ground, nursery trees should have lateral branches removed and be pruned and topped to a single whip at approximately 24 inches above ground level.

One-year-old whips (single-stemmed nursery stock) with many active buds are ideal. Occasionally, two-year-old trees are used, and in that case the trees are again topped at approximately 24 inches, but lateral stems are stubbed back to a couple of buds to give the tree growing points at the desired height of scaffold limb development.

A common practice is to wrap the young tree with aluminum foil from slightly above ground level to approximately 16 inches above the ground. The aluminum foil suppresses shoot growth below the desired level for scaffold limbs, can protect tender trunks from sunscald should weather turn abnormally hot in the spring, and can protect young trees from incidental contact from postemergence herbicides. Grow tubes are also commonly used for this purpose and accomplish the same goal. Grow tubes are commercially manufactured sleeves of plastic or other material sold to protect trunks of young trees. We allow for a single trunk up to 18 inches because as trees get older, it will be much easier to manage competing vegetation on the orchard

▲ After planting, a one-year whip should have lateral growth removed and be topped at about 24 inches.

▶ If nursery stock is older than one year at planting, stub back lateral growth to provide growing points at the desired height.

floor than if the crotch were lower to the ground. A common mistake among new growers is to treat trees as miniatures and develop scaffold systems closer to the ground. Once topped, the trunk will not grow any taller; in other words, once the crotch of the tree is developed at a given height, it will remain at that height for the life of the tree.

First Growing Season

The general school of thought on training first leaf (first growing season) trees is to try to establish scaffold limbs with a minimal amount of hard summer pruning in order to capture as much sunlight as possible. Remember, summer pruning is a dwarfing action, so be cautious how much green tissue you remove in striving to develop scaffold limbs. During the first growing season, suckers or any shoots forcing below the height at which scaffolds are desired should be removed as soon as they are noticed. It is critical that these shoots be removed as close to the tree as possible. If these shoots are not removed at the point of attachment with the scaffold limb, they will grow back in larger numbers with a vengeance. With ample water, nitrogen, and weed control, first leaf trees should all produce sufficient growth to select a permanent set of scaffold limbs. When selecting for scaffold limbs, try to divide equally the imaginary circle one sees when looking down on the tree from above. Resist the urge to leave more than four primary scaffold limbs. While more limbs may make the trees appear fuller when they are young, more than four scaffold limbs will crowd the developing tree, and inevitably, one or more will have to be removed in order to overcome the shading of other limbs. By the end of the first summer the trees should begin to take on the normal open vase appearance. As summer turns into fall, the green shoots lignify (turn from succulent to woody) and turn a mahogany brown as trees begin to acclimate for the coming winter.

First Dormant Pruning

The goal of the first dormant pruning is to select and establish the permanent scaffold system on each tree. Many new growers have a hard time severely pruning the young trees they worked so carefully to grow—but remember, dormant pruning is an invigorating action.

Excessively vigorous shoots called water sprouts should be removed because they are a source of shade for the desirable limbs retained after pruning. Select three or four vigorous limbs spaced apart as equally as possible and remove all other shoots from the trunk.

Avoid selecting limbs that will shade each other as well as limbs that are excessively upright or horizontal. Limbs that have narrow crotch angles should be avoided too—approximately 45 degrees is the desired angle in relation to the trunk for permanent scaffold limbs. These scaffold limbs should be pruned back to approximately 24 inches for formation of sub-scaffold limbs in the second growing season. When possible, primary scaffold limbs should be pruned back to a position where two strong shoots form a "Y" and can provide the structure necessary for the formation of secondary scaffold limbs in the second growing season

Occasionally, shoots arise from below ground called root-sprouts or suckers. These shoots should be removed and pruned back as close to the trunk or rootstock as possible. Simply cutting them at ground level will result in regrowth and a persistent problem with suckering for years to come.

The goal of the first dormant pruning is to establish scaffold and sub-scaffold limbs that will be retained as permanent parts of the tree.

Second Growing Season

Depending on the level of vigor, some summer pruning may be required during the second growing season. Remove any shoots growing downward on the primary or secondary scaffold limbs you retained during dormant pruning. It is also wise to walk through the orchard several times during May and June to remove suckers and water sprouts while they are still small. This will increase the overall vigor of the tree and allow the remaining growth to initiate flower buds for a partial crop in the third leaf.

Second Dormant Pruning

As always, suckers, any new growth from the trunk, and water sprouts are removed from the tree first. Pruning during the second winter should be used to ensure that any missing scaffold limbs are now in place and to finalize selection of sub-scaffold limbs 18 to 24 inches from the crotch of the tree. At this point, upright and downward-hanging annual growth is removed in favor of fruiting wood growing laterally within the canopy. Downward-growing fruit wood can be left

Dormant two-year-old trees before and after pruning.

if there are no better choices in a portion of the tree, but limbs should be stubbed back so that they do not bend to the ground with the weight of the fruit. Low-hanging limbs and fruit interfere with weed control, and fruit touching the ground will be highly prone to rotting or fire ant infestation. The key is to learn the proper balance of pruning form and function. While it is important to try to maintain the shape of the tree, overthinning of fruit wood will cause a reduction in crop in the following growing season.

Subsequent Pruning

Water and nitrogen promote vigor in all horticultural crops. Summer pruning may be needed more in years with high rainfall or in years of crop loss that lead to excessively vigorous growth. Remember, summer pruning is most effective when practiced early in the summer and in moderation. To affect flower bud initiation in peach trees, summer pruning should be finished by the end of June so that remaining desirable green shoots have optimal sunlight interception for a sufficient crop of flower buds to be formed over the remainder of the summer.

Dormant pruning needs to be conducted annually over the life of the orchard. Failure to do so will result in proliferation of growth in the interior of the tree, which will shade and ultimately kill desirable limbs in the preferred fruiting zone. Regenerating neglected trees can be accomplished but typically comes at the cost of that year's crop. A severe dormant pruning to invigorate the tree and regain the proper shape is needed, followed by frequent summer pruning trips to direct growth in the desired location within the canopy.

In a well-maintained orchard, annual dormant pruning becomes structurally less severe than in early years, but vigorous trees still demand a considerable amount of time to prune. Trees should be topped, preferably at about 8 feet, but no higher than 9 or 10 feet in height. With the weight of the crop on the limbs, topping at this height will result in all fruit being harvestable without ladders. Topping also removes apical dominance and allows for greater lateral spread of the tree. The general growth habit of "up and out" creates a martini glass shape, and a mature orchard has a dense canopy that casts shade on all the ground underneath the trees down the tree row. Centers should remain open, and in some cases the work needed for removal of

Even for an experienced grower, a mature tree may take as long as 40 minutes to prune. Mature trees (top) are usually pruned each year by removing vigorous growth in the interior of the tree (center), and then making heading cuts to keep the tree growing up and out, and thinning out fruiting wood (bottom).

water sprouts is substantial. Maintaining the tree form, keeping centers open, and thinning out the fruiting wood for the approaching crop constitute the annual chore of pruning.

Mechanical Aids

All growers look for ways to save labor in the orchard, and the option of pruning trees mechanically has been investigated for at least four decades. Mechanization of the annual chore of pruning mature orchards typically results in canopies with either increasing amounts of shade or excessive removal of fruit wood. The result is either poor cropping levels or poor fruit quality. That said, many growers have utilized mechanical tree toppers to help in the time-consuming chore of maintaining tree height. Hydraulic powered sickle-style cutter bars, somewhat similar to old hay cutters, can be used to top trees at a given height. Fitted with heavy duty blades, they can typically cut through wood as great as an inch in diameter. Using this type of mechanical assistance can reduce the pruning labor by many hours per acre. Mechanical pruning, however, may increase the probability of spreading bacterial canker in infected orchards.

▲ Peach trees need a dormant pruning each year. If not pruned, growth concentrates in the interior of trees, resulting in shaded, dying limbs in the desirable canopy and a dense interior canopy that is conducive to fungal pathogens. Note the brown rot mummies in the interior of the tree.

▶ Some larger growers use mechanical aids such as sickle-bar hedgers to reduce the amount of hand labor needed for dormant pruning.

Chapter 7

. . . .

Water Management

O ne of the most critical aspects of peach culture, be it one tree or several acres of trees, is water management. Our philosophy has always been that one has to get the water right in order to be successful. Water is the medium for all the physiological processes of the plant and hence is critical not only for optimum but also for maximum growth. Times when water is most critical to the plant include planting, young tree growth, mature tree bud break and bloom, fruit sizing and crop set for the next year, fruit maturation, and the fall food storage time for the tree.

Planting

Water is critical to ensure the survival of the newly planted tree. Even if the soil is wet, the tree must be watered in at planting with 5 to 7 gallons of water. This water removes the air pockets from the planting hole and puts the soil in intimate contact with the roots. A drip irrigation system will *not* do this job adequately. Once the tree has been planted and watered in, the hole has been refilled with some soil to compensate for air pockets, and the tree has been rewatered to settle the new soil, the tree should not be rewatered for about six weeks. The tree is planted during the dormant season and is using virtually no water. New root growth needs oxygen to flourish, and a weekly water application after planting removes this oxygen from the planting hole. Once the tree breaks dormancy in the spring, then a weekly water application is in order.

Young Tree Growth

The objective of young tree growth is to grow a large tree in two years so that some fruit production is possible in the third year. Fabulous growth the first year often leads to some fruit production in year two. One-gallon drip emitters are ideal for establishing young trees. In year one each tree needs one emitter; in year two it needs three emitters; and in years three to fifteen it needs six to ten emitters or one micro-sprinkler.

Peach tree irrigation takes into account rainfall and water applied and furthermore assumes immaculate weed control around the trees. Newly planted trees are poor competitors for water and nutrients, and if weeds are allowed to grow around the trees, maximum tree growth cannot be achieved. Assuming that adequate rainfall has been received after tree planting, it is typically not necessary to begin water applications before April. However, if it has been extremely dry, it will be necessary to begin water applications six weeks after planting. The following table gives monthly and weekly applications that have proven successful.

If the soil fails to drain properly and/or if water stands around the trees it may be necessary to reduce the volume of water applied. These numbers are merely guidelines. A system that has proven successful in determining irrigation timing is to feel the soil. Dig down 3 inches and feel the soil. If the soil "balls together" there is still adequate water in the soil. If the soil does not ball but crumbles, then water should be applied.

Mature Tree Bud Break

Most irrigation systems are supplemental at best; this is especially evident in dry years. Mature peach trees, with a canopy of 18 to 25 feet, use up to 100 gallons of water per day in July and August. However, these are tree water requirements and not irrigation requirements.

Water required per tree per week, by month (gallons per tree per week).

Tree Age	March	April	May	June	July	August	September
1	7	7	7	14	28	28	14
2	14	14	14	28	56	56	28

Peach tree roots have the potential to exploit the entire soil profile for water. We consider the effective root zone of a peach tree to be about 3 feet. In some areas, tree roots may have 8 to 10 feet of soil to mine, whereas others may have only 10 to 12 inches of soil and then solid rock. The better the soil, the better the trees will do, despite any weakness in the management program. We hope for sufficient rain in January, February, and early March to replenish the root zone of the trees totally. In this way, the trees have the potential for a strong bloom and bud break if proper chilling is received. If water is lacking at bud break, bloom can be weak and sporadic. Stressed blooms are less likely to set. Once fruit is set, it seems to grow for about six to eight weeks and then noticeable growth stops.

Fruit Sizing and Flower Bud Initiation

After a lull in growth the fruit begins a rapid increase in size, which culminates in fruit maturation. This early sizing, followed by a lull and then rapid growth, is described as a sigmoid growth curve (shown in chapter 12). Water is a definite must once this rapid sizing begins, if full potential genetic size is to be achieved. Remember, all the cells for the fruit are present when the fruit is set, and water provides the turgor pressure for these cells to achieve their full size. It is critical that adequate water is present at this time to ensure large fruit. At this same time, the tree is also initiating the flowers for next year's crop. Adequate water is a must to ensure the formation of normal flowers. Undue stress to the tree during this period will cause the formation of twins or double fruit the next year.

Fruit Maturation

The largest increase in the size of the fruit occurs in the last two weeks before maturation. Adequate water during this final time prior to harvest is needed to achieve large fruit. Some growers have the notion that too much water just before ripening reduces the overall flavor of the fruit. But keep in mind that the leaves produce the sugars that eventually accumulate in the fruit just before harvest. This sugar production occurs in photosynthesis, which is at its maximum when the sun is shining. If a rainy period occurs prior to ripening, sunshine is limited, and so is sugar production—hence the lack of flavor in the fruit. Consistent, uniform application of water up until harvest will ensure large and flavorful fruit.

Food Storage

Once the tree has matured its crop, its water requirement declines. However, the tree's water need does not go to zero. Water is needed to maintain tree photosynthesis and physiological processes as well as to store carbohydrates for cold hardiness. Stressing the tree by withholding water is erroneously thought of as hardening off the tree, but this is far from the truth. Stressing the tree reduces its ability to produce the carbohydrates it needs to go dormant in good condition. We therefore maintain a regular but reduced water schedule into the fall to ensure proper plant dormancy.

Application Methods

Now that we know the times when water is critical for optimal growth and fruit quality, we can detail water application methods.

Even though drip irrigation is ideal for peach production, water berms around the young tree work well. The key with both systems is that as the tree grows, the number of emitters must increase and/or the size of the berm around the tree must increase. The most effective absorbing roots begin at the drip line (canopy edge) and extend out; if the soil is extremely shallow, the roots will go out more than double the height of the tree. Hence this is where the water must be placed. Also, the best absorbing roots are in the top 6 to 12 inches of the soil but may extend to 3 feet down in really good soil. However, the zone of available water should never be allowed to go below 3 feet, as the tree must expend energy to capture this deep water. This energy could be better used to size the fruit or store food for next year's crop.

An ideal water situation for mature peach trees is an inch of water per week. If we have 100 trees per acre, this amounts to about 270 gallons of water per tree per week, or about 39 gallons of water per tree per day. The ideal drip system would apply this amount of water every day. A mature peach tree would typically have six 1-gallon-per-hour emitters; hence the system would run a little over 6 hours

Drip irrigation is an ideal way to establish peach trees and it works well on large trees.

Irrigation risers are placed in mid-row with lateral lines going in either direction to assist with consistency in water pressure within the irrigation system.

Water berms work well on newly planted trees.

every day. Some growers choose to apply the water on a weekly basis, which is okay as long as the water does not percolate below the 3-foot zone. Coarse soils do not have the same water holding capacity as heavier soils. In general heavy soils should be watered longer and less frequently, and coarse soils should be watered for shorter periods of time but more frequently. Another thing to consider is that the tree harvests water from its entire root zone. Hence we can assume that in a normal rainfall year the tree is getting water from soil areas that are not wet by the irrigation system. It would be only in very dry years that the full complement of water would need to be applied.

Irrigation System Design

If ample water is available (10 gallons per minute per acre of trees planted), and your venture is directed at being a viable commercial operation, then you should have your system designed to provide the maximum amount of water required on a daily basis, or about 39 gallons per tree. Realize that in rainy years your irrigation requirement will be less because the tree will harvest part of its daily 39-gallon water requirement from the soil reservoir. Still in dry years, not only does your fruit size suffer, but your tree health is compromised as well.

System design can be simple if you have just a few trees but can be complicated if planting several acres. We encourage letting the professionals design the system. In this way pipe size and resulting pressures will be appropriate, and the resulting water output across the orchard will be uniform. Several irrigation supply houses will design your system free of charge if you purchase the materials from them. These design people need to know your well or water capacity and water quality in order to design the system properly. Some elect to put the system in by themselves; others settle for a turnkey job. Local garden centers and box stores carry a wide range of drip irrigation supplies for those who have just a few trees and want to install an irrigation system themselves.

Regardless of system used—drippers or micro-sprinklers—the system must be turned on prior to the trees needing water. Roots concentrate in wet areas if given time to grow there, as roots constantly die and regenerate. However, if you turn on your micro irrigation system for the first time when the trees need water, there will be only limited roots in the area where the water is being applied. Hence the initial uptake of this water is going to be limited. Once this area is wet for a while (7 to 10 days), roots will congregate there.

Ideally drip irrigation emitters are operated for 8 to 10 or perhaps 12 to 16 hours at a time. In this way the water does not go below 3 feet, and the off period allows oxygen to reenter the root zone. Little if any water can be absorbed from saturated soil, as oxygen is necessary for the uptake of water. Typically, micro-sprinklers have to be operated for a longer period of time as they wet a larger volume of soil and so it takes longer to push the water 6 to 12 inches deep. Evaporation losses are higher with micro-sprinklers and the sprinklers have to be checked religiously to make sure they are working.

Again regardless of system used, one must have a way to check the soil moisture. One cannot assume that a tree is getting water just because the faucet is in the "on" position. Emitters and micro-sprinklers must be checked regularly to make sure they are not plugged. In addition, the soil-wetting pattern must be checked after the stated run time has elapsed to see how deep the water advanced. Perhaps the system was run too long or possibly not long enough.

Scheduling Irrigation for Mature Trees

Some growers water on a daily basis; that is, they replace the amount of water the trees use every day. This amount will vary based on the

Micro-sprinklers work well on large trees.

Wetting pattern of the soil.

environmental conditions. Hot and windy weather promotes greater water use than on cloudy and cool days. Others apply their irrigation water once a week; typically this is an inch of water per week. Another possibility is to measure the soil moisture. Numerous tensiometers and/or watermark sensors are available; however, the feel of the soil is probably your best bet. You dig down 3 to 4 inches and get a handful of soil. If the soil balls together, it still has ample moisture; if the soil crumbles and does not hold together, water needs to be applied.

Water Filtration

The heart of the irrigation system is the filtration unit. If you are using a sprinkler on the end of a hose, it is not a big issue. However, if you have several acres of trees and six emitters or one micro-sprinkler per tree, filtration does become a big issue. The small orifices on both the emitters and sprinklers are prone to clogging.

Small amounts of sand, silt, and clay can be removed with a sand media filter down to about 70 micrometers in size. A settling pond is needed if the water is heavily contaminated with silt and clay. Very small particles (less than 70 micrometers) will typically pass through the system without incident, but these particles settle out over time. Hence it may be necessary to chlorinate the system. Chlorination of 10 to 20 parts per million (ppm) two to three times during the run

A handful of soil that balls together means the soil has adequate moisture.

The filtration unit is the key to keeping a drip system operating properly.

Clogging is typically not an issue with large sprinklers.

cycle is usually sufficient. It would be advantageous to allow the chlorination to remain in the lines once the run has been completed.

Iron, sulfur, and calcium carbonate can be especially troublesome. Iron is usually present in the underground water source in a soluble or ferrous form. Unfortunately when the water enters the irrigation system, iron is often precipitated as insoluble ferric oxide, which can clog the emitters. The iron can be removed from the system with aeration and settling: oxygen precipitates the iron as ferric oxide, and it falls to the bottom of the tank. Chlorination likewise causes this precipitation to occur, and then the precipitate can be removed with a sand media filter. Chlorination will also control sulfur precipitation. Last, iron is more soluble in a low pH solution. Hence the system can be slightly acidified to maintain the iron solubility, or the pH in the lines can be dropped to 4 occasionally to dissolve the iron and flush it out of the system. This acid treatment will also take care of calcium salts, which may precipitate in the lines. Thus not only is filtration important in the operation and maintenance of micro irrigation systems, but the injection of acid and chlorine is important as well. Such injection devices also allow for the injection of fertilizer, which is the most efficient way to apply fertilizer.

Fertilizer Injection

Fertilizer can be applied effectively through the drip irrigation system. Soil wetting is limited using drip irrigation, and dry fertilizer applied on the soil surface may not be effectively moved into the root zone. Connect the fertilizer injector into the main pipeline so that fertilizer can be selectively routed to all trees in the orchard. To prevent emitter plugging, fertilizer must be injected upstream from the filter so that all undissolved fertilizer material will be removed by the filter.

An electrically driven positive displacement pump, a water-motor proportioner, or a venturi proportioner can be installed as an integral component of the drip irrigation system. Select an injector that operates properly on the electrical voltage, water pressure, and water flow rate available in the orchard. An injector with capacity to supply adequate fertilizer for large trees should be selected initially.

Liquid nitrogen such as Uran can normally be applied by the irrigation system. Check for formation of a precipitate that can clog emitters and filters by mixing small amounts of the liquid fertilizer material with irrigation water prior to injection. Test all fertilizers, including proposed fertilizer mixtures, before injection into the irrigation system. Often iron is required on high pH soils, and an iron chelate can be effectively injected through the system.

Begin adding fertilizer when the irrigation system is applying water at the normal rate. Check the application rate by timing the injection of a specific quantity of material. Complete injection before the irrigation cycle ends, so that most fertilizer material will be moved into the root zone and the irrigation system will be flushed. One to three hours may be required to move fertilizer material to trees at the end of lateral pipelines, especially where only one emitter is used for each tree.

Calcium and/or sand can clog emitters.

Soil Salinity

All water from streams and underground sources contains dissolved materials known chemically as salts. Generally, "salt" is thought of as table salt, but many different salts are known. In discussions of irrigation water quality and soil salinity, the many different salts found in the water are referred to collectively as "salt."

Most irrigation water does not contain enough salt to be injurious to plants. It is important to test the water thoroughly prior to using it for irrigation. Let the well run for one hour prior to taking the sample. Then collect the water sample in a disposable plastic baby bottle. Ship the sample overnight to the lab. The principal effect of salinity is to reduce the availability of water to plants; however, certain salts or ions may produce specific toxic effects.

Salts accumulate in the soil around the edges of the wetted pattern under drip irrigation emitters, and some leaching (removal of salts with drainage water) may be required. Sufficient rainfall is received in much of the state to accomplish most required leaching of salts. However, addition of extra irrigation water may be required in some areas to leach accumulated salts from the root zone. Operation of the irrigation system when the water requirement of the trees is low can probably accomplish required leaching of salts in most cases. When the irrigation water contains significant quantities of salt, it is advisable to conduct an annual salinity analysis of soil samples from the root zone as well as of the irrigation water itself. Ideally the water should have the following qualities:

- Sodium absorption ratio (SAR) should be less than 3
- Total salinity should be less than 1 millimhos per centimeter (mmho/cm)
- Bicarbonate should be less than 2.5 milliequivalents per liter (meq/l)
- Less than 500 ppm sodium (Na)
- Less than 1 ppm Boron (B)

Operating Drip Irrigation Systems

Orchard water requirements are influenced by tree size and growth stage as well as temperature, relative humidity, and wind velocity. Irrigation time can be controlled manually, with a time clock. Inspect the orchard regularly to determine the possible need for adjustment in daily irrigation time.

A final word—plant only the number of trees that you can irrigate properly or that your water supply can handle. Overambitious planting can lead to a failed venture. If you do a poor job of watering, the trees will eventually let you know, but by then they will be stressed to the point of injury.

Chapter 8

. . . .

Peach Orchard Floor Management

Weed competition is a leading cause for failure of young peach orchards. Up to 90 percent of the tree roots are in the top 2 feet of soil, and weeds growing in the root area of the tree are in direct competition with the tree for water and nutrients. In young orchards, uncontrolled weed height also results in competition for sunlight. Peach trees are poor competitors, and the grower's ability to manage competing vegetation has a tremendous impact on the success of the orchard. Growers should avoid the temptation of mowing only the row centers, or aisles, and ignoring weed growth down the tree row underneath the trees. While it is much easier to mow than to deal with weeds under the trees, vegetation from trunk to drip line is the most important to control.

Growers face the challenge of needing to remove weed competition but at the same time to ensure that valuable and irreplaceable topsoil is not lost. The recommended method of managing the orchard floor is to develop a weed-free strip under the tree row with a permanent sod strip between rows. Mechanical tillage of the entire orchard floor was the standard of the past but is rapidly losing favor because of root pruning and erosion. Sod middles in properly contoured rows can stop or at least greatly reduce erosion, but they should be mown frequently to reduce insect and disease pressure. Sod middles also provide good footing for the movement of orchard equipment after a rain, which can be important in harvesting or making timely fungicide applications.

Tillage

Tillage has long been used to control weeds in orchards and is still used in many production regions today. Yet tillage, especially deep tillage, results in the loss of many shallow feeder roots. The concept is that tillage essentially turns the tilled 2 to 4 inches of topsoil into mulch and forces the root system to develop below the tillage depth. In very deep sandy soils where air penetrates deep into the soil profile, this approach may be workable; but in heavy soils, tillage can seriously reduce the vigor of mature, bearing trees. Roots are important not

Poor weed control is the main reason new orchards fail.

only for water and nutrient uptake but also as vessels of carbohydrate storage and as regulator of growth and reproductive physiology.

One more reason why tillage is no longer recommended is that it results in soil erosion. Soils take thousands of years to create and should be considered an important, irreplaceable resource. It has been documented that a deeply tilled orchard can lose as much as 2 feet of soil in a single high rainfall season. These are catastrophic events that can render a site useless for future production. Further, it is a misconception that tillage is the environmentally friendly approach to weed control. The environmental impact of burning fossil fuels many times

Optimal floor management consists of a weed-free strip under the trees and tightly mown aisles.

In addition to pruning vital root tissue, tillage can promote the loss of valuable topsoil through erosion.

a season could arguably be the most unsustainable option available to fruit growers.

Access issues arise too. It is a fact of life that commercial peach production requires a grower to make timely, sometimes critical fungicide applications to control fruit and foliage diseases. When an orchard is tilled and a high rainfall event occurs, the orchard floor may be impassible for equipment for a time. A delay of several days before a tractor and sprayer can enter the orchard may prove disastrous and result in high fruit loss to fungal or bacterial diseases. Repeatedly driving up and down a tilled row can also develop what is known as a plow pan, a compacted zone that restricts the penetration of roots into the soil. Finally, repeated tillage negatively impacts soil structure and can reduce root growth and water holding capacity.

Mulch

Peach growers may consider the use of organic mulch to manage weeds. To be effective, mulch should be applied at least 6 inches deep. In our hot southern climate organic matter dissipates quickly in or on the soil, so applications will indeed need to be made annually to be an effective means of managing competing vegetation. Straw, spoiled hay, and leaf litter are all appropriate materials for this purpose. Some growers have chosen to use ground tree trimmings, which are frequently available from maintenance crews of utility companies. While in the long run this source of organic matter will break down and be beneficial, it can initially compete with peach trees for nitrogen used by microbes to break down the chips.

The benefits of mulch to the orchard are numerous. First, mulch can effectively inhibit weed growth, keeping water and nutrients from being wasted by competitive vegetation. Second, mulch is effective in further conserving moisture by preventing evaporation from covered soil under the trees. The use of mulch and the continual breakdown of organic matter also has several positive impacts on soil structure, can help neutralize soil pH, and can increase the availability of specific micronutrients.

Mulching is not frequently employed in larger commercial orchards because of the cost of both the mulch material and the labor needed to deliver the mulch in place.

Chemical Weed Control

Chemical weed control is cost competitive with mechanical tillage because chemical control requires less costly equipment, fewer trips through the orchard, less horsepower, and less fuel. With chemical weed control, the only orchard floor equipment really needed consists of a herbicide sprayer and a mower. To use herbicides safely and effectively, the commercial grower should have a basic understanding of the types of chemicals and application methods available.

Postemergence Herbicides

Contact Herbicides

These are simply postemergence chemicals that kill vegetative tissue on contact. These herbicides provide quick knockdown of green top growth on most weeds. Some well-established annuals (plants that grow from seed each year) and most perennials (plants that live for several years) will regrow and may require frequent applications to keep weed growth under control.

Gramoxone, marketed as Paraquat, is a contact herbicide commonly used in and labeled for peach orchards. Paraquat is very effective on weeds and safe around peach trees as long as it does not directly contact green leaves or young tender bark. *Paraquat is highly toxic to humans and should be handled very carefully.* Paraquat has no residual soil activity.

Systemic herbicides are postemergence chemicals that penetrate plant tissue and are moved throughout the plant and into the roots. These chemicals usually kill slowly, but they kill the entire plant, from the roots up.

Glyphosate and related compounds include Roundup (glyphosate), which is labeled for use in bearing and nonbearing peach orchards. Glyphosate kills a broad spectrum of weeds and grasses and is particularly valuable because of its effectiveness against highly competitive perennials such as Bermuda grass and Johnson grass. Glyphosate is much more effective when broadleaf weeds and grasses are in a succulent stage of growth. Weeds under drought stress are somewhat tolerant of injury by glyphosate. Glyphosate has no residual soil activity.

Certain perennials such as perennial morning glory, nightshades, dewberry, poison ivy, and perennial ragweed are relatively tolerant of glyphosate. The best results with glyphosate on difficult to kill weeds are obtained with fall applications in late September, October, or early November at least several days before a killing frost. More food materials are being translocated to plant roots in the fall, and proportionately more glyphosate will move to the roots as well and provide greater efficacy.

A number of generic glyphosate products and compounds related to glyphosate are currently on the market. These products are toxic to peach trees and care must be taken to avoid direct contact or contact from spray drift on leaves or green bark. Spray applications should be made when there is little to no wind, and young trees should be shielded to prevent incidental contact.

Selective systemic herbicides currently labeled for peach orchards include two products that will kill annual and perennial grasses but not harm broadleaf plants, including peach trees. Fusilade (fluazifop-butyl) is labeled for bearing and nonbearing peach orchards, and Poast (sethoxydim) is labeled for use on nonbearing peaches. The chief advantage of these materials is around small peach trees where grasses have gotten out of control and it is impossible to avoid contact of the spray material with the trees. In order for chemicals such as these to be effective, grasses must be in a succulent, young stage of growth. Specific crop oil surfactants are recommended with these products, and two applications may be required for adequate control.

Preemergence Herbicides

Preemergence herbicides are those which are applied to a portion of the orchard floor to inhibit the germination of annual broadleaf and grassy weeds. In order for these materials to be effective, they must be uniformly applied to the orchard floor and incorporated to a shallow depth by mechanical cultivation or rainfall before weed germination.

With some preemergence materials, organic debris should either be closely mown or be incorporated into the soil before application. A heavy cover of sod or plant residue may prevent much of the chemical from reaching the soil, resulting in reduced effectiveness. The most effective times to apply preemergence herbicides are in February, prior to germination of most spring weeds, and/or in September, before germination of most winter weeds.

No single preemergence herbicide is effective against all weeds. The choice of herbicide(s) should be based on the weeds and grasses present in the orchard. Application rates of some preemergence herbicides are based on length of desired control, while for others rates are based on soil texture. Consult the most current *Southeastern Peach, Nectarine and Plum Pest Management and Culture Guide* for specific product information, rates, and restrictions. Total elimination of weeds is seldom achieved with preemergence herbicides. Successive annual applications will gradually reduce problems of "escape weeds," but periodic spot treatments with a postemergence chemical are usually necessary under most orchard conditions.

Methods of Herbicide Application

Boom Sprayers

Sprayers with a tractor-mounted or sprayer-attached boom are useful for of all types of liquid-applied herbicides. Because this equipment provides for agitation of spray material in the tank and is easily calibrated to deliver precise spray volume, tractor-mounted sprayers are the only type of equipment suitable for preemergence herbicides. A typical boom

Tractor-mounted boom sprayers greatly increase weed control efficiency.

is mounted about 18 inches above the ground and fitted with 80-degree flat-fan spray tips spaced 20 inches apart. This spray boom configuration with 80-degree tips along the boom axis delivers overlapping spray that result in a uniform curtain of spray material on the soil. These sprayers do not require high pressure or high volume for spray delivery. A typical herbicide sprayer will not need to deliver more than 50 gallons of solution per acre and will not need to generate pressure of more than 30 psi.

Backpack (Knapsack) Sprayers

These types of sprayers are ideal for spot treatments of weeds with postemergence herbicides. A good operator can also accurately apply preemergence herbicides with a backpack sprayer if able to maintain consistent pressure and steady walking speed. These sprayers are better for applying liquid chemicals than wettable powders since they have no means of solution agitation other than shaking.

Backpack sprayers offer tremendous advantages over handheld pump-up sprayers. Handheld sprayers can be used to apply postemergence herbicides, but because of variability of sprayer pressure and applicator ground speed, they should never be used for preemergence materials.

Controlled Droplet Application

Controlled droplet application (CDA) sprayers have spinning heads, which disperse the spray solution as fine, uniform droplets. They are available as handheld units or with boom-mounted heads for tractor application. CDA is used primarily with postemergence herbicides, especially glyphosate. It offers advantages in chemical savings and in applications with a minimum volume of water. Handheld units are used successfully by some peach growers to apply glyphosate even in large orchards.

Wick Application

Wick applicators or weed wipers are used primarily as a safe method of applying glyphosate where sprays cannot be used safely, such as close to small trees or under windy conditions. Wick application of glyphosate works best on tall, more easily controlled species such as Johnson grass. Lower growing and tougher to kill grasses such as Bermuda grass are controlled more effectively and efficiently by spray application of glyphosate, assuming it can be applied safely.

Calibrating a Herbicide Sprayer

Accurate calibration of the herbicide spray rig is essential. Application of too little chemical is ineffective and wasteful of the herbicide since an additional application will have to be made to get the desired results. Application of too much herbicide can be both wasteful and dangerous to the trees. Overdoses of certain preemergence herbicides

Herbicide boom calibration can be accomplished with a tape, a volumetric container, a watch, and simple math.

can be especially hazardous. It is important to remember that while insecticides and fungicides are recommended in amount of product per acre of orchard, herbicides are recommended in amount of product per *treated* acre.

The applicator must know how much spray solution is being applied to each treated acre in order to determine the amount of chemical to add to the tank. Numerous methods can be used to calibrate sprayers. One of the simplest is presented below. Calibration steps shown are for a tractor-mounted boom, but the same steps can be used to calibrate a backpack or CDA sprayer.

Simplified Herbicide Calibration

1. Select the appropriate herbicide(s) and rates for the expected weed spectrum and soil type present.
2. Thoroughly clean out spray tank, lines, and spray boom.
3. Partially fill spray tank with clean water, start pump, and set to desired pressure. Check spray nozzles for even pattern and uniform delivery. Measure width of spray band.
4. Measure off a test area (e.g., 100 linear feet). With the tractor operating at desired ground speed and the spray pump operating at desired pressure, time how long it takes equipment to travel over the test area.

5. With the equipment stationary, run the pump at the same rpm and pressure, and measure the volume of water that is delivered from each of the nozzles in the same time it took to travel the test area. Because of potential variability between nozzles, use a measuring cup to measure each nozzle separately. The volume from each should be the same; if it is not, consider replacing worn nozzles. Total the volume from all nozzles. This will give you a water volume to area ratio that will allow you to calculate the amount of water volume being applied per treated acre.

Example: Assume the boom covers a 5-foot swath and it takes you 20 seconds to travel the 100-foot test area. If there are four nozzles and each puts out 15 ounces in 20 seconds, there are 60 ounces of water being applied over a 500-square-foot area. A simple ratio allows you to calculate volume per treated acre.

$$\frac{60 \text{ ounces per acre}}{500 \text{ sq. ft.}} \quad = \quad \frac{?}{43,560 \text{ (\# of sq. ft. per acre)}}$$

$60 \times 43,560 = 2,613,600$; and $\dfrac{2,613,600}{500} = 5,227.2$ ounces per acre

$$\frac{5,227}{128 \text{ (\# of ounces per gallon)}} \quad = \quad 40.8 \text{ gallons per acre}$$

You then simply put the appropriate amount of herbicide(s) per treated acre in approximately 40 gallons of water.

Cover Crops

The maintenance of vegetation in row centers is vital for the long-term health and productivity of peach trees. While in some locations and in some situations legumes are valuable as a pre-plant cover crop, they only fix nitrogen when nitrogen is deficient in the soil. Under those conditions, nitrogen will be insufficient to grow healthy trees. Most growers choose simply to utilize existing native vegetation for a spring and summer cover rather than to plant a specific crop, such as

Cover crops improve soil structure, prevent soil compaction, improve equipment mobility, and help prevent soil erosion.

clover or vetch. Row centers are typically maintained through mowing as needed throughout the season. Keeping the aisles mown will help optimize air movement through the orchard.

Winter covers are commonly broadcast or drilled into row centers. Elbon rye, Austrian Winter Pea, or more commonly annual rye grass is used to increase organic matter in the orchard. A winter cover crop can also help keep an orchard floor cool, delaying bud break in the spring. A word of caution, however: the cooling effect of cover crops can be a detriment during a frost. A tall-statured cover crop inhibits an orchard floor from absorbing heat during the day, and the increased surface area quickly dissipates any heat the floor has absorbed at night.

Cover crops can either be continually mown until heat kills the winter cover, or herbicides can be used to knock the cover crop down at bloom. The use of annual rye as a winter cover has an additional advantage. When rye grass is grown and then killed with herbicide, its decaying roots inhibit the germination of other weeds. This feature may make the choice of annual rye grass appropriate for use under the trees as well as in row centers.

Peach Tree Nutrition and Orchard Fertilization

n order to maintain healthy, productive trees, the fertility of peach trees must be kept at optimal levels. The diversity of soils and climates in Texas makes it impossible to provide a blanket recommendation for fertilization from one end of the state to the other. For this reason, we must rely on soil and leaf tissue samples in order to discern the most cost-effective amount of nutrients to be applied on an annual basis. The timing and placement of fertilizer also affects tree growth and productivity. As growers, our goal should be to produce a vigorous canopy early in the growing season, to reduce vegetative vigor near fruit maturity, and then to maintain healthy leaves late into the fall.

Soil samples should be taken prior to planting so that if the pH needs to be raised, lime can be applied. Elements that move slowly in the soil, such as phosphorous, may also be added and incorporated into the orchard site before the trees are planted. Yearly monitoring of soil and leaf nutrient levels helps determine if adjustments need to be made in the orchard's fertilization program. In some cases, leaf tissue samples point out the need for a nutrient for the following year's crop, and fertilizers may be added in the fall so that they have time to reach the root zone of the trees. Summer nutrient monitoring may also reveal a need for minor nitrogen fertilizer applications late in the season so that the canopy does not senesce prematurely. Maintaining a healthy canopy as late in the fall as possible will improve the winter hardiness of trees. A small, late fall application of nitrogen to bearing orchards is commonly done so that some nitrogen absorption can occur with the typical flush of root growth in the fall and through the winter.

The concept of nutrient mobility is important when diagnosing nutritional disorders in the orchard. Mobile elements are those that the plant has the ability to extract from older tissue and move to the growing tip. Consequently, deficiencies of mobile nutrients such as nitrogen and magnesium are first seen in older leaves. Immobile nutrients are those which cannot be moved within the plant. Deficiency symptoms of immobile nutrients are first seen in new growth.

Soil Types Impact Fertilization Practices

Soil particles contain exchange sites that help soils retain certain nutrients. These exchange sites are *negatively* charged, so nutrients that are positively charged (cations), such as phosphorous, magnesium, and potassium, become held to these exchange sites. This helps hold these nutrients in the soil and make them resistant to loss due to leaching as a result of rainfall or excessive irrigation. The movement of some cations through the soil may be fairly rapid, such as magnesium, while others may be very, very slow, such as phosphorous.

Nutrients such as nitrogen are typically found in the soil as negatively charged compounds (anions), which are not bound on soil exchange sites and consequently are subject to movement in the soil. High rainfall can force mobile nutrients such as nitrogen out of the root zone of peach trees, and the fertilizer is subsequently lost to leaching. Consequently, strategies for applying anions and cations may be very different.

A tree's ability to absorb positively charged nutrients is dependent on:

* Nutrient availability, which may be limited by soil pH
* Adequate water in the soil so that nutrients may be extracted
* The presence and concentration of other nutrients that may compete for uptake

Nitrogen (N)

More than for any other element, nitrogen management presents a dilemma to new and seasoned peach growers alike. A grower must manage this element in order to balance vegetative growth and fruit production effectively. Because trees are very responsive to nitrogen fertilization, it is all too easy to tip the scales in either direction. In almost every soil type, nitrogen applications are needed annually to encourage strong, vigorous growth, but soil types vary in their ability to hold nitrogen. In very coarse, sandy soils where water moves freely, nitrogen may best be applied in many small doses. Nitrogen fertilizers come in several forms and usually include nitrate, urea, and ammonium sources of nitrogen. Nitrate is the form of nitrogen most readily utilized by peach trees, but the nitrate ion is subject to several fates in the soil profile. Urea and ammonium sources are broken down by soil microbes and are converted into nitrate ions. With adequate water and oxygen in the root zone, new root tips readily absorb nitrogen. In either dry or waterlogged soils, peach trees often exhibit nutritional deficiencies that are a direct result of limiting soil conditions. When excessive moisture is present in the orchard, nitrate ions are leached out of the soil profile when water drains. In saturated soils, nitrate ions can be chemically converted to nitrite and can be lost through volatilization.

Nitrogen is generally recommended in units of actual nitrogen per acre. To convert this number to fertilizer rates, divide by the percent nitrogen of the fertilizer material. For example, an application

Foliage well supplied with nitrogen will have a dark green color.

of 100 pounds of ammonium nitrate (33% N) would result in the application of 33 units of nitrogen; 100 pounds of ammonium sulfate (21% N) will provide 21 units of nitrogen.

For young trees, small, frequent amounts of nitrogen provide the best conditions for strong growth and scaffold limb establishment. Monthly applications of nitrogen fertilizers at the rate of about one half cup per tree from March until July or August is generally sufficient to supply the annual nitrogen need. In newly planted orchards, nitrogen fertilizers should be placed in a ring approximately 18 inches from the trunk. Growers should be cautious using nitrate forms of nitrogen because over-application or placement too close to the trunk of the tree can cause serious injury or death to trees.

Research in the southeastern states has shown that 45 to 90 pounds of nitrogen per year is generally adequate for bearing orchards. Applications should be split between late bloom and late June. In many cases, fall applications of nitrogen help retain leaves in a healthy condition late into the fall. A rule of thumb on the amount of fall nitrogen is 15 pounds/acre on trees exhibiting healthy foliage and little or no terminal growth and 30 pounds/acre on trees with no growth and a light green color. Trees vigorously growing in late August should not receive the postharvest application of nitrogen.

Nitrogen Deficiency Symptoms

Nitrogen is a mobile element, so deficiency symptoms are first seen as the yellowing of older, basal leaves. In severe cases, shoot tips appear yellowish green and older leaves appear red or brown and shed prematurely.

Symptoms of Excess Nitrogen

Excess nitrogen results in delayed ripening, decreased red color of fruit, and excessive vegetative growth. In addition, excessive nitrogen or placement of nitrogen fertilizers near the trunks of young trees can cause severe root burning and even tree death.

Phosphorus (P)

Phosphorus deficiencies are rare and applications should always be made on the basis of soil and leaf tests. Peaches remove only 12 to 15 pounds/acre of phosphorus annually, so high application rates should be avoided. Because phosphorous moves extremely slowly through the soil profile, sites with low soil phosphorous should be treated before planting so that the fertilizer can be incorporated down into the root zone.

Excessive Phosphorus

Many Texas soils test high in phosphorus. The continued application of phosphorus to these soils commonly results in zinc, iron, and/or copper deficiencies. This applies especially in the alkaline soils of Central and West Texas, where iron chlorosis is a problem. Phosphorus is also the most expensive of the three primary macronutrients, so phosphorous fertilization should be avoided unless leaf analysis shows a deficiency.

Potassium (K)

Potassium has been shown to influence fruit color and cold hardiness in peaches. Both trees and fruit buds have been shown to be more cold hardy when adequate potassium levels are maintained. Color and size can suffer when leaf levels are below 1.25 percent. Since potassium leaches readily into the subsoil, soils with a long history of potassium fertilization should have a subsoil analysis to determine if an accumulation has occurred. If subsoil levels are high and leaf tissue analysis indicates potassium in the adequate range, additional potassium fertilization would not be warranted.

Deficiency Symptoms

Potassium-deficient trees develop light green to pale yellow leaves rolling inward in a bean pod shape. Severe symptoms include tip necrosis and a crinkled midrib. Fruit bud development is reduced in deficient trees.

Potassium fertilization should be based on soil and leaf analysis due to the wide variability in potassium content of Texas soils. Where soil levels are low, annual applications of 70 to 100 pounds/acre of K2O equivalent may be needed to maintain adequate potassium levels.

Iron (Fe)

Iron deficiency is common in those areas of Texas with alkaline soils (pH over 7.0). Iron deficiency can be a severe problem where the pH is above 7.5. The symptoms are an interveinal chlorosis of new growth, where veins remain green but interveinal tissue becomes yellow, white, or necrotic. Contributing factors to iron deficiency are excess moisture and/or poor soil drainage, extremely high soil temperature, nematodes, and high calcium bicarbonate. Excess phosphorus fertilization also exacerbates the chlorosis problem. In many of these soils, soil sample results will show adequate iron. Iron is chemically bound and made unavailable to many plants in high pH soils. Adding amendments such as iron sulfate is generally ineffective because the iron is quickly bound in the soil. To a degree, the negative effects of high pH soils can be corrected through deep mulching, but this practice is costly and only partially effective. Rootstock selection can also greatly impact the extent to which trees exhibit iron chlorosis.

In many cases the use of chelated iron products is the best practical means of dealing with iron chlorosis. Chelated nutrients are those coupled with organic compounds that resist chemical binding in the soil. The plant absorbs the chelated molecule and extracts the needed

Chlorotic terminal tissue in this high density peach planting is indicative of iron deficiency.

element. Iron chelate is most effective when applied to the orchard floor at the drip line of trees. While effective, this treatment is expensive. Chelated materials have also been used effectively as foliar sprays. Although these products will regreen chlorotic growing tips, iron is an immobile nutrient and may require frequent applications to supply iron adequately to new growth. Applications of 1 pound per 100 gallons of spray solution are sufficient, and iron chelate products are tank-compatible with many fungicides and insecticides. Check pesticide labels for specific restrictions. Chelated iron products are formulated for either acid, iron-poor soils or alkaline soils where iron availability is a problem. Growers should make sure they select the correct product for a particular orchard site.

Zinc (Zn)

Zinc deficiency is most commonly found in alkaline soils and in those acid to neutral sandy soils with very low levels of zinc. Symptoms

Zinc deficiencies can be seen with chlorotic, narrow foliage and a proliferation of lateral bud growth.

are chlorotic, mottled, narrow crinkled new leaves. As the deficiency worsens, the twigs appear stunted with a rosette of leaves near the terminals.

In acid soils low in zinc, deficiencies can be partially corrected with the application of fungicides high in zinc. Soil application of 1 pound of zinc sulfate per tree spread under the drip line of the trees has also given positive results. Zinc chelate sprays can help in the correction of zinc deficiency as well. Zinc is another immobile nutrient, so product application must begin very early in the growing season to be effective.

Sulfur (S)

In recent years sulfur has been identified as a limiting factor in some of the acid soils of East Texas. Soil and leaf analysis for sulfur are recommended for growers on acid sands to determine if this is a problem. If a deficiency exists, it can readily be corrected by applications of sulfur-bearing fertilizers. Sulfur-deficient trees appear light green and resemble the early stages of nitrogen deficiency. The lower leaves, however, do not turn yellow as with nitrogen deficiencies.

Soil and Tissue Testing

Mid-July is a good time to collect both soil and leaf samples. Accurate records should be maintained so that the nutritional analyses from each block of trees can be compared easily from year to year.

The following table shows nutritional ranges in peach leaves to help you understand your leaf analysis results.

Soil samples are divided between topsoil and subsoil samples. Both should be taken prior to planting so that the entire nutritional picture is known to the new grower. Topsoil samples should consist of random slices of soil from the top 8 inches of the soil profile. Exclude rocks and organic matter. The procedure for proper soil sample collection is printed on the soil analysis information sheets available from each county AgriLife Extension Service office. The accuracy of both soil and leaf samples are dependent on the collection of a representative sample. If problem areas are observed, they should be sampled separately from the rest of the block.

When sampling blocks for nutritional analysis, samples should

Nutritional ranges in peach leaves.

Element	Deficient	Hidden Hunger*	Sufficiency
N (%)	< 2.0	2.1–2.7	2.75–4.00
P (%)	< 0.11	0.11	0.12–0.50
K (%)	< 1.00	1.10–1.49	1.50–2.50
Ca (%)	< 1.0	1.10–1.49	1.50–2.50
Mg (%)	< 0.20		0.20–0.50
Mn (%)	< 20		60–400
Fe (ppm)	< 20		60–400
B (ppm)	< 20		20–100
Cu (ppm)	< 3		5–20
Zn (ppm)	< 12		15–50
S (ppm)	< 1,000		1,200–1,500

*Hidden Hunger is that range where no symptoms are evident, but the tree will respond to the addition of this element.

consist of about fifty leaves: two leaves collected from each of twenty-five trees selected randomly from a given block. Remember that different varieties must be sampled separately. Choose representative shoots on bearing limbs from the perimeter of the tree that have had good sunlight exposure, and select leaves midway down the shoot. It is strongly suggested that leaves selected for nutritional analysis be washed before they are analyzed. Briefly (10 seconds) submerge and gently rub all of the leaves from a given sample in a mild dish soap/water solution (phosphorus-free is best). Then rinse them two or three times, using distilled water for at least the last rinse. This removes from the surface of the leaves soil and fungicide residues, which can result in spurious readings. After rinsing, place the leaves on a paper towel or in an open paper bag in a warm, well-ventilated area. Allow the leaves to dry completely (become crisp) before sending them to the lab for analysis. Place leaves in a paper bag (do not use plastic bags). Paper sandwich bags are ideal. Label each sample bag with at least your name and block identification number, and include a sheet listing the identity of all of your samples. Leaf samples may be submitted using a Forage and Tissue Testing Form, available at any county extension office.

Chapter 10

. . . .

Peach Diseases

Peach orchards are subject to numerous disease organisms that can decay fruit or cause defoliation or other serious injury to all or parts of trees. In our climate, even in drier parts of the state, organic production of peaches is unprofitable commercially. Diseases of foliage and fruit are generally driven by rainfall and humidity and are therefore more problematic in wetter parts of the state. This chapter is not intended to be an exhaustive discussion of disease biology and epidemiology but merely an overview of some of the more problematic diseases across the region. One in a series of important references, *Compendium of Stone Fruit Diseases* (edited by J. Ogawa, E. Zehr, G. Bird, D. Ritchie, K. Uriu, and J. Uyemoto and published by the American Phytopathological Society, 1995) is the single fullest reference for growers looking for more substantial detail on peach diseases. Because the registration of disease control products changes constantly, we have avoided discussion of specific products and rates. Instead this discussion centers on understanding the biology of the most common diseases and presenting cultural strategies that are important in disease management.

Diseases of Foliage and Fruit

Brown Rot

The single most destructive disease affecting peach fruit around the world is brown rot, caused by the fungal pathogen *Monolinia fructicola*. During the spring, with sufficient inoculum and high rainfall, brown

▲ Brown rot mummies can be retained in the tree canopy over the winter and give rise to disease outbreaks the following year. Care should be taken to remove mummies from the orchard during pruning.

▶ Brown rot–infected fruit begin to dehydrate and show spore-producing structures that can lead to higher disease severity.

rot can infect and destroy blossoms and young vegetative shoots. Once fruit are set, they remain susceptible to infection through the duration of the ripening process. Infection on ripening fruit can be seen by a rapidly spreading brown lesion that eventually engulfs the entire fruit. When left in the tree, infected fruit eventually dehydrate and remain affixed to limbs. These "mummies" are a major source of inoculums for subsequent infections.

Latent fruit infections can be highly problematic after harvest and can dramatically shorten the shelf life of fruit. In storage, a single infected fruit can rapidly deteriorate, produce spores, and give rise to infection of adjacent fruit in the same container or storage facility.

Brown rot overwinters as mummies in trees or on the ground and in infected vegetative tissue. Removal of blighted shoots and mummified fruit is an integral part of overall disease control. Conidia, which give rise to spores for subsequent infection, can be formed on previously infected tissue at temperatures of 41°F and higher. In terms of the impact of weather on infection, it really does not matter how much it rains; the issue is how long the canopy of a tree remains wet. With as little as three to five hours of canopy wetness and temperatures of 65–70°F, significant infection can take place. Infection can occur at any temperature from 41° to 86°F and is most pronounced between 65° and 70°F.

Commercial management of brown rot is dependent on protective fungicides, which are typically applied several times during the growing season. The frequency of application depends on the amount of inoculum in the orchard and the frequency and duration of wetness in the canopy. Orchard sanitation can facilitate disease management by reducing inoculum, and canopy management through dormant and summer pruning can help reduce the duration of canopy wetness following precipitation. In our state, commercial organic production of peaches is at best very difficult. Homeowners looking to manage brown rot with organic materials such as sulfur will be dependent on sanitation, canopy management, and luck to ripen a crop. With frequent rainfall, there will be significant losses, but these techniques may prove acceptable in drier years. Neither author recommends commercial organic peach production in any area of the state. Under high rainfall seasons, management of brown rot can be challenging even with the frequent application of commercial fungicides.

Peach Scab

The second most important fungal disease of fruit in the South is peach scab, caused by the fungal pathogen *Cladosporium carpophilum*. Scab is common on peaches and nectarines in warm, humid growing areas of the world and is sometimes found on plums. Leaves and shoots can become infected, but symptoms are more noticeable on fruit.

Mild peach scab infection results in a spotted cosmetic blemish that is inconsequential for home consumption but is not acceptable in a wholesale setting.

Fruit infection from peach scab usually takes place relatively early in the season, but symptoms are first seen when the fruit is approximately half grown. Symptoms can range from superficial spots on the peach skin to sunken lesions causing the cracking of skin when multiple lesions coalesce. Lesions on shoots appear at about the same time as fruit infection becomes evident but is almost unnoticeable until later in the season, when slightly raised oval lesions appear on the current year's shoots.

These stem lesions produce new spores in the spring about two weeks before shuck split (see last photograph in phenology box, chapter 4). Sporulation on twigs occurs with high humidity, and germination is optimized at or near 100 percent relative humidity. In essence, rainfall from petal fall until approximately 30 days postbloom constitutes the period that fruit are susceptible to peach scab infection. Control of peach scab is entirely dependent on the application of fungicides. Because vegetative lesions are numerous and hard to discern, there are no practical sanitation practices that can aid in the management of peach scab. Fungicide applications are currently recommended from petal fall through shuck split for control of scab. Frequency of application is dependent on the frequency and duration of canopy wetness. For homeowners, mild infection is entirely cosmetic and may be ignored, but visible scab infection is problematic for commercial (especially wholesale) growers. Severe infection that leads to skin cracking can increase fruit susceptibility to brown rot.

Bacterial Spot

This disease is caused by the bacterial pathogen *Xanthomonas campestris* pathovar *pruni*. The pathogen can infect stems, leaves, and fruit, causing economic losses because of crop loss and/or loss of leaf area. Foliar infection first appears as water-soaked spots along leaf tips and midribs, which later become necrotic. Lesions can become purple or brown and may become necrotic and result in a shot-hole appearance. Severe infection can cause significant to complete defoliation, which can result in crop loss or a reduction of fruit quality. This loss of vegetative area can severely stress trees, resulting in decreased winter hardiness and overall poor tree health.

Varieties differ in their resistance or susceptibility to bacterial spot. Growers in high rainfall areas should pay careful attention to the sus-

Bacterial spot can cause shot-hole lesions in foliage and sunken lesions on the surface of fruit. Severe infection can lead to defoliation of trees.

ceptibility of specific fruiting varieties before making planting decisions. In high rainfall years that result in a dense canopy, summer pruning in the interior of the tree can help reduce drying time after a rain and aid in mitigating bacterial leaf spot pressure. Control of bacterial spot is largely dependent on dormant application of copper hydroxide compounds. In the past these sprays were recommended in late fall, but recommendations have changed in recent years. Copper defoliates trees, and our need to maximize postharvest photosynthesis has resulted in our now recommending complete leaf drop before application. The pathogen overwinters at the leaf scar (where the leaf was attached to the shoot), and waiting until complete leaf drop allows us to cover our target areas better with dormant sprays. These compounds act by neutralizing the pathogen, thereby reducing disease inoculums. In winters following seasons with severe infection, growers should consider two copper applications—one at 100 percent leaf fall and the second at the first sign of bud swell in very early spring. In severe situations, antibacterial pesticides may be needed, but they are expensive and under scrutiny for renewed federal pesticide registration.

Peach Leaf Curl

In cool to cold climates with prolonged cool, wet springs, peach leaf curl can cause significant damage, but in our climate this disease is of little commercial significance. Caused by the fungal pathogen *Taphrina deformans*, this disease results in grotesquely malformed foliage;

leaves become reddened and puckered and commonly abscise from the tree. The disease appears early in the spring with the full expansion of the first leaves, but in our warm climate it usually stops infecting new tissue shortly after it first appears. Leaf curl is controlled with the dormant application of copper hydroxide. Because of the widespread use of these products to control bacterial spot, peach leaf curl is now an infrequent curiosity that does not require further control measures.

Peach Rust

Peach rust is widespread across growing regions, but the frequency and timing of the onset of symptoms is regulated by temperature and rainfall during the growing season. Although there are two species of rust found across the world, only *Tranzschelia discolor* has been documented in the southeastern United States. Symptoms appear as pale yellowish green spots on the underside of leaves, although rust can be present on both upper and lower leaf surfaces. These spots become raised and mature into orange to tan slightly raised lesions. Shoots can also become infected, but symptoms generally go undetected. While

While seldom economically problematic in our climate, peach leaf curl causes distinct malformation of foliage.

infection can occur earlier in the growing season, peach rust is usually seen as an infrequent disease of late summer through early autumn.

When infection is widespread, it can lead to early defoliation, resulting in lower tree winter hardiness. In some growing regions, rust can infect fruit, resulting in water-soaked-appearing green lesions that become leathery and stand out as blush begins on ripening fruit. While foliar rust infection can occur with some regularity in higher rainfall areas, there is little if any history of fruit infection in Texas, and rust is not considered a significant threat to fruit. In seasons with frequent rainfall, fungicide applications may be needed to retain foliage.

Diseases of Woody Tissue and Roots

Bacterial Canker

One of the most widespread diseases across peach orchards in Texas, bacterial canker can be severe and can lead to decline and loss of major scaffold limbs and ultimately to tree death. Although there are several species of bacterial canker organisms that infect *Prunus* species across the world, bacterial canker of peach in the southern United States is thought to be caused solely by the pathogen *Pseudomonas syringae* pathovar *syringae*. The organism overwinters as limb cankers

Airblast sprayers are commonly employed to spray foliage and fruit in commercial peach orchards.

Bacterial canker can cause sap to run from trunks or limbs and is commonly misidentified as borer damage.

and can reside systemically in tissue that otherwise appears healthy. In the spring, bacterial canker can invade leaf tissue and open pruning wounds, causing shoot and limb collapse. As the season progresses, especially as conditions turn hot and dry, the tree can display an outpouring of sappy gummosis that emanates from infected tissue. In many parts of the state this symptom is commonly misdiagnosed as injury from wood-boring insects.

Systemic canker infection at these suspect sites can be confirmed by peeling back bark where black longitudinal streaking can be found. While trees can live for quite some time with bacterial canker, it is highly advised that growers follow through on cultural practices that can reduce disease spread and maintain tree health. Extended periods of drought appear to predispose trees to bacterial canker infection.

Bacterial canker usually enters trees through some type of mechanical wound. One of the primary ways *Pseudomonas* finds an entryway into peach trees is through freeze injury. Freeze injury is commonly not seen until one or two years after the cold event and in many cases can first be identified by the presence of new canker infections. Trees that are otherwise predisposed to freeze injury are by definition weakened, and their ability to ward off infection is greatly reduced compared to that of healthy trees. Pruning cuts (especially large ones), deer rutting, and injury from tillage or other orchard equipment can provide entry sites for the organism, resulting in disease spread. Some growers have reported a perceived increase in bacterial canker from summer pruning in wet years. Even chewing insects such as grasshoppers can move the bacteria from primary or supplemental hosts in and around the orchard, causing new infections.

Controlling bacterial canker in the orchard is dependent on a mul-

Post oak root rot causes trees to decline and finally die. The fruiting structure of these fungi can occasionally be seen at the base of infected trees.

tipronged management strategy. Dormant applications of copper hydroxide made for bacterial spot and peach leaf curl control are also effective in neutralizing bacterial canker inoculum and help minimize new disease incidence. Delaying pruning until late winter and painting large cuts with pruning paint can result in substantial reduction of new disease incidence within an orchard. If canker infections are observed, pruning out infected limbs and retraining young trees with suitable replacement limbs may be necessary. Pruning out known infected tissue in older trees is also advisable. In either case, removal of infected pruning trash from the orchard and burning or burying it away from other trees is advisable.

It is also important to sterilize saws, loppers, and other pruning equipment after cutting through tissue known or suspected to be infected. Denatured ethanol is ideal but may be somewhat expensive or unavailable. A 5 percent sodium hypochlorite (Clorox) solution can also be used but can quickly rust pruning equipment. Frequent application of light oil may be needed to maintain the good working condition of all treated equipment. Taking precautions outlined in discussion of dormancy (chapter 4) to maximize tree health and cold hardiness can provide significant assistance in helping a plant ward off new infection or remain productive in spite of infection.

Post Oak (Mushroom) Root Rot

The primary soil-borne pathogen affecting peach trees in the central and eastern portions of Texas and elsewhere in the southeastern United States is *Armillaria mellea*. A second organism, *Clitocybe tabescens*, causes similar symptoms and tree death patterns but can be discerned by subtle differences in the white fungal mat at or near ground level when it occurs. For all intents and purposes, these organisms cause the same disease problems and the difference in causal organisms does not affect disease incidence or management.

Post oak root rot can cause severe tree losses and commonly begins appearing in orchards after about five years. It can be common on orchard sites that have been recently cleared of timber, especially hardwood forests. While newly cleared sites are at greater risk of disease incidence, the disease can persist on dead and decaying roots for many years. The disease is more common on sandy coarse soils as opposed to sites with greater clay content. Symptoms on peach trees usually begin in midsummer, when all or parts of a tree begin to wilt, usually from the apical tip back toward basal tissue; foliage usually yellows, followed by defoliation and a rapid collapse of the entire tree. Trees that become infected late in the season may be very slow to begin growth in the spring and may have small cupped foliage prior to tree collapse. Extremely dry or excessively wet soils can also make peach trees more susceptible to infection. Post oak root rot usually spreads within an orchard by the extension of mycelia strands along root tissue or perhaps even by direct growth through the soil. Because of the close proximity of roots, orchards planted at high densities are more subject to tree-to-tree spread than are trees with older dry-land orchard spacing. Infection sites usually spread in a circular or oval-like pattern outward from the center of infection.

Other than avoiding sites prone to post oak root rot, there is no known control of this disease. Soil fumigation is futile because the disease can persist well below where the soil can be penetrated by volatilized chemicals, and no fungicides are known to be effective against this disease. Although *Prunus* species vary in their susceptibility to *Armillaria*, all present commercially available rootstocks that are graft-compatible with peach are susceptible to this disease. New rootstocks utilizing related *Prunus* species with higher tolerance to this disease complex may in time provide commercially acceptable alternatives to the current susceptible rootstock choices. In replant sites or on sites with a known presence of *Armillaria*, it is suggested that the soil be worked and as many roots from old plants be removed as possible. Planting a replant site for two or more seasons with a highly productive cover crop and incorporating that crop into the soil are recommended to create a healthy soil microbial population that may compete against the development of this root rot complex and help prevent it.

Cotton Root Rot

Although it has been documented in slightly acidic soils, this pathogen is only problematic in high pH soils in Central, South and West Texas. Cotton root rot is not known to exist north of the Caprock in the High Plains region of Texas. Recently renamed *Phymatotrichopsis omnivora*, this fungal pathogen has daunted growers of many crops across alkaline soil areas of Texas, southern Arizona and New Mexico, and much of northern Mexico. Contrary to popular belief, the pathogen is not isolated to sites previously planted in cotton. The species name *omnivora* means "eats everything" because there are more than 2,500 cultivated plants on its known host range. While nowhere near as sensitive as apples, peaches are indeed susceptible to cotton root rot. Peach tree losses to this pathogen are common where it has been heavily established or in orchards under stress from insufficient or excessive soil moisture.

No fungicides have proven to be effective in preventing infection by or losses from cotton root rot in perennial crops. Growers have made many attempts to address the threat of this pathogen by adding sulfur or other amendments to lower the pH of the soil. While it is a common effective practice to raise soil pH with the addition of lime, it is virtually impossible to lower soil pH effectively with acidifying agents. Invariably, such efforts drastically drop the soil pH in the top few inches of soil, but soil below the cultivation line is unchanged. The high exchange capacity from calcium bicarbonate in the soil is most commonly the reason such efforts are unsuccessful. There is simply too much bicarbonate and calcium in the soil for even high amounts

Cotton root rot death is usually very sudden. Sudden tree collapse with retained foliage is symptomatic of this disease.

of sulfur to neutralize. Numerous products are promoted and sold as biological solutions and solutions for cotton root rot. While the concept of a healthy, diverse soil microbial population competing with this pathogen in the soil makes sense, the problem lies in our inability to deliver these products to the deepest roots of the trees. The products may have benefits to shallow-rooted annual crops, but they have not been demonstrated to have value in inhibiting losses in peach orchards. Although there appear to be some differences in susceptibility to cotton root rot among rootstocks, all commercially available stocks otherwise adapted to our area should be considered susceptible. Evaluation of experimental *Prunus* hybrid selections in Texas shows that there may be some promising rootstock alternatives available in the future.

Other Peach Diseases

Phony Peach

This disease is caused by a specific pathovar of the bacterium *Xylella fastidiosa* that produces stunting of trees and malformation of fruit. Infected trees have shortened internodes (distance between leaves on a stem), and the foliage usually appears dense and dark green. Although the name was supposed to be "pony peach" (because of the short stature of infected trees), a typographical error in initial reporting resulted in the permanence of the name phony peach. The bacterium resides

benignly in numerous other native plants, where it is acquired by several
species of insects known as sharpshooters. The bacterium colonizes the
foregut of the insect and is transmitted to peach through sharpshooter
feeding on green shoots with their needlelike mouthparts. Epidemics
have been experienced twice in the twentieth century across the South-
east (including East Texas) and appear to coincide with warm winters
that favor survival of the pathogen. There is no cure for phony peach,
and the only mechanism of control is to remove infected trees from the
orchard to prevent further disease spread.

Plum Pox Virus

A devastating disease in much of Europe, plum pox (also known as
sharka) was not known in North America until it was found in Can-
ada in the 1990s. This disease, which is spread through infected nursery stock and by aphids, results in distorted foliage and fruit. A survey
conducted by USDA revealed some infected trees in Pennsylvania,
resulting in a quarantine and destruction of infected orchards. The
survey included Texas and showed no other infested areas.

Crown Gall

The bacterium *Agrobacterium tumifaciens* infects and causes symptoms on numerous woody perennial plants. Infection can usually be
detected by the formation of galls on roots or the crown of the plant
at ground level. On occasion, galls can be found on aerial portions of
peach trees. At first the galls appear as disfigured tan or brown growth
that grows rapidly, ultimately producing a dark brown to blackish gall
as much as 4 or 5 inches in diameter. Infection arises either from contaminated nursery stock or from naturally occurring bacteria entering a wound near ground level. While crown gall can cause serious
economic losses on some crops, such as grapes, the effects seem to
be much less severe on peach trees. Crown gall infections are more
common on alkaline poorly drained soils but can be found on trees
planted in any soil type. Mechanical removal of galls or topical treatments that reduce gall size have been employed as a control strategy
but appear to have no economic advantage. Studies conducted on infected versus noninfected trees show no reduction in growth, yield, or
fruit quality as a result of this disease.

Chapter 11

. . . .

Insects

.

umerous insects can hinder both tree life and fruit quality. Fortunately, only a few pose a serious threat to the trees and fruit in most years. The number one insect responsible for peach tree mortality is scale, followed by greater peach tree borer and lesser peach tree borer. Fruit pests include the stinkbug complex, plum curculio, and oriental fruit moth. Luckily many areas of Texas do not have to deal with curculio, and borers do not occur in the Texas Hill Country. Mites, thrips, and grasshoppers can be an issue in some years.

Scale

Scale insects have been a serious problem and destructive pest on peach trees all across the United States. The most damaging species to peach trees is probably San Jose scale. Scale are tiny, orange, saclike insects beneath light gray waxy coverings and with rasping/sucking mouth parts. Each scale covering is made up of a series of concentric rings, surrounding a raised nipple near the center. Females give birth to living young called crawlers. These tiny young insects crawl from beneath the parent scale to suitable places on the bark, leaves, or fruit and insert their mouthparts. After feeding for a few days, the young scale secrete their waxy scale coverings.

Unfortunately scale insects can persist in all stages of development throughout the year. Mature scale and even nymphs can survive even during cold weather. Scale populations can drastically increase fol-

San Jose scale on the tree and fruit.

Insects

White peach scale is common in Texas.

lowing harvest, especially in wet summers. The heavy foliage and lush twigs provide ideal conditions for scale development. Due to their small size and obscure coloration, scale insects can reach damaging levels before growers realize they have a problem. The reproductive rate of San Jose scale can be exceptionally high; luckily dry summers can help hold the pest in check.

A grower who fails to detect high scale populations begins to see a decline in the tree's vigor, followed by sparse, yellow foliage as the scale literally suck the life out of the tree. Reddish spots may be found on the underside of the infested bark as well as on the leaves and fruit.

Another scale pest of peach trees, not as destructive as San Jose scale, is white peach scale. In severe infestations of white scale, they appear as white, cottony masses encrusting the bark of the tree.

Due to the destructive nature of both kinds of scale insects as well as the possible infestation of several species of soft scale, annual applications of dormant oil are advised on all peach trees. Even if scale is not detected, one should make an annual dormant oil application to control undetectable scale numbers before they become problematic. An added benefit of such a spray is that it helps control aphids and mites as well. The dormant oil spray should be made as close to bud break as possible. If one has a severe scale infestation, two applications may be in order. Over the years many insecticides have changed and are no longer available; however, the application of dormant oil remains a constant in the quest to grow quality peach trees and fruit.

Peach Tree Borer

Further damage to the trees can occur from peach tree borers. Many consider peach tree borer to be the most destructive pest of mature peach trees, though scale has killed more trees. A large mass of gum that contains sawdustlike material near the soil line on the trunk of the tree is a sure sign of borer damage. To be sure that indeed you have borers, you need to cut under the gum area to see if a worm or larva is present.

If indeed you have borers, you will find the larva and its track. The larva feeding activity can be so extensive that it girdles the tree. Females lay eggs on the lower trunk of the trees. This egg lay can extend from early May through October. Generally the eggs hatch in about ten days and the tiny larvae burrow directly into the trunk. They can

The larval stage of the peach tree borer causes damage to the tree and is visible in gum exudates.

111

Insects

feed on the tree all winter, and larvae pupate in early April with a single generation per year. It is advisable to look for evidence of borer damage in late spring to early summer. If borers are found, annual applications of insecticide targeting this pest are advisable for the life of the orchard.

Lesser Peach Tree Borer

The lesser peach tree borer (LPTB) is quite similar to the peach tree borer in morphology and life history. However, the larva prefers to feed in the trunk and limbs rather than at the base of the tree. Infestations tend to occur in crotches, under old bark, and in cracks or openings produced by physical injury. Diseased trees and those with freeze damage or sunscald tend to be preferred hosts for the LPTB. Usually only one and a partial second generation occur. Overwintered larvae change to pupae when warm weather arrives. Moths emerge from the pupae and lay their eggs on the trunk and limbs. Newly hatched larvae select weak points on the tree and bore into the cambium. This feeding of the LPTB in the wood weakens the tree and limbs and provides entry for rot-producing organisms and shot-hole borers.

Proper care of trees is important in maintaining an orchard free of LPTB. Poorly fertilized, drought stricken, and diseased trees are highly susceptible to attack. LPTB are seldom an issue in well-managed orchards, but care should be taken to prevent mechanical damage to trees. Lesser peach tree borers are attracted to damaged areas in the bark, and careless pruning can provide favorable egg-laying sites around the stubs of branches. Dead and broken limbs should be removed and destroyed, and trees should be maintained in the best possible condition year-round to reduce the chance of having this pest.

Stinkbug Complex

Several species of stinkbugs and plant bugs cause a gnarling and distortion of the fruit called catfacing. Basically these pests have a needle for a mouthpart, and when they feed on the fruit to suck out the sap, they kill the cells at the point of feeding. If the fruit do not abort as a result of this attack, fruit development is inhibited in the area of the punctures, while the surrounding healthy tissue continues to grow.

Various species of plant bugs can be found in peach trees during the pink bud stage. Damaged buds, blossoms, and small fruit often fall following their feeding activities. If the damaged fruit remain on the tree they will develop sunken, corky areas. The bug populations usually decline after petal fall, as the pests are attracted to other hosts.

Small green stinkbugs are usually the first bugs to cause economic damage to the peach crop. They emerge from their overwintering sites (leaf trash, tall grass) and fly to peach trees from the late bloom stage until after shuck split. Generally they migrate to other hosts shortly thereafter. Damage from green stinkbugs can be severe, and by harvest time, the fruit that have fed on are folded and distorted.

On the other hand, brown stinkbugs are in the trees shortly after petal fall. The adults appear in largest numbers about a month after shuck split and typically remain in the trees throughout the season. Fruit damaged early develops corky, depressed areas, whereas fruit fed on late is only slightly damaged.

Various species of stink bugs and their damage to the fruit.

Southern green and green stinkbugs are generally the last species to appear in damaging numbers. Peaches on trees along orchard edges or woodlands are the most severely damaged. Feeding damage is different from that of other stinkbugs, as the damaged fruit appear water soaked and dimpled. Frequently gum exudes from the feeding punctures.

Prevention of injury from catfacing insects largely depends on sanitation around the trees or orchard. Leaves and tall grass and weeds should not be allowed to accumulate around the trees, and unimproved pastures adjacent to peach plantings should be mown to remove supplemental feeding and reproductive sites for stinkbugs. Sprays may be required when the pests are present in large numbers.

Plum Curculio

The plum curculio is probably the most challenging fruit-attacking pest to control. Fortunately, it is not present in many areas where peaches are grown. Woodlands, fence rows, and ground trash provide overwintering sites for the plum curculio. Typically this overwintering generation prefers plums and/or nectarines as egg-laying sites, though they lay on peaches as well. The snout beetle adults feed on the fruit and deposit their eggs at these feeding sites. Such holes provide entry points for brown rot as well.

After placing eggs on the fruit, the female cuts a crescent-shaped slit under each egg. As the larvae hatch they feed and tunnel in the fruit, which often drops from the tree. If the fruit do not drop, when these larvae are full-grown they drop to the soil and pupate. A new crop of adults emerges in about four weeks and the cycle begins again. The larvae from this second generation are often found feeding in the fruit at harvest time. Adult plum curculios continue to feed on the foliage until cooler weather arrives, at which time they migrate to overwintering sites such as ground trash.

Plum curculio is revealed by crescent cuts on the fruit after egg laying, and the larva inside the fruit makes it unsalable.

If this pest is present in your location, it will be necessary to be on a calendar spray program to prevent damage. Four spray applications are common for this pest: at shuck split, two additional applications at two-week intervals, and usually a final spray about thirty days before harvest. Obviously, early-maturing varieties would need fewer spray applications.

Oriental Fruit Moth

An important pest of late-maturing peach varieties is the oriental fruit moth. Spring feeding injury is usually confined to the growing shoot tips. As a result some shoot terminals can die. Larvae of succeeding generations feed on the fruit and often enter through the stem. They also gain entrance where fruit are side by side. Infested fruit break down rapidly because of extensive internal damage. The oriental fruit moth passes the winter in the larval stage in a cocoon under the bark, in mummified fruit, or in ground trash.

Injury is typically severe in orchards that also have apples, pears, and late maturing peaches. The use of varieties that ripen before mid-July decreases the chance of severe damage by this pest.

Mites

Mites can cause a serious defoliation issue in some years. Mites prefer dusty, hot conditions, and some pesticides can cause them to flare. Carefully monitor your trees for leaf loss caused by this pest. Avoid the use of several harsh sprays like Sevin or pyrethroids in a row, as such sprays can kill beneficial insects and allow mites to get out of control.

Mites have a very short generation time and with some miticides, two applications seven to ten days apart are needed for control. This pest commonly slips up on growers in late summer after harvest, and mites can defoliate an orchard in very short order. Keeping an eye on your orchard after harvest until first frost will pay dividends in keeping trees healthy.

Thrips

Some highly colored, early-ripening peach varieties can be attacked by thrips one to two weeks before harvest. Feeding by these tiny insects

Early recognition of mite infestation requires inspection of the underside of leaves to see webbing and feeding injury. Left untreated, mites can lead to defoliation of peach trees very rapidly.

results in the appearance of silvery scraped areas on the fruit surface. Though the skin is never ruptured, the fruit is unattractive. Small amounts of cosmetic damage are generally tolerated in the marketplace, but extensive damage may not pass inspection with wholesale or retail buyers.

Grasshoppers

Young peach trees can be attacked voraciously during dry summers. The insects prefer the tender bark and fruit skin to the foliage. Often time's twigs are girdled due to their feeding. Normally, grasshoppers only attack trees when other crops are spent in summer. Inspect weeds around the edges of young orchards during the spring for small grasshoppers. If large numbers of nymphs are observed it may be necessary to treat this peripheral area. Orchard treatment is usually unnecessary if the grasshoppers are controlled at the edges of the orchard.

Thrips cause superficial silvering on the surface of fruit skin.

Control recommendations for homeowners can be found in the *Homeowner's Guide to Pests of Peaches, Plums and Pecans* (by Allen Knutson, Kevin Ong, James Kamas, Bill Ree, and Dale Mott, published by Texas Cooperative Extension Service, 2005), online at http://www.plantanswers.com/homeowner_peach_guide.pdf.

Grasshoppers feed on both leaves and fruit.

Key to Insects Attacking the Peach Tree and Fruit

Insects Feeding Externally on the Fruit

Green or brown stink bugs, or plant bugs sucking the sap from fruits, and/or sap oozing from the fruit
—**catfacing insects**

Tiny, elongated yellow insects feeding on shaded portions of the fruit, causing scraped, silvery areas
—**thrips**

Green or brown insects eating shallow irregular depressions on the surface of ripening fruit
—**grasshoppers**

Larvae Feeding inside the Fruit

White, legless larvae with brown heads feeding in the fruit
—**plum curculio**

Pink-white, active caterpillars with abdominal legs burrowing in the fruit
—**oriental fruit moth**

Insects Attacking the Trunk and Large Limbs

White caterpillars feeding in upper trunk and large branches
—**lesser peach tree borer**

Larvae tunneling in the lower trunk
—**peach tree borer**

Stationary Insects on the Trunk and Branches

Light-gray waxy scales with raised nipples in the center
—**San Jose scale**

Bark covered with a white, cottony mass of scales with yellow to orange centers
—**white peach scale**

Insects Feeding on the Foliage and Twigs

Insects feeding on the leaves and tender bark of small trees, frequently girdling branches

—**grasshoppers**

. . . .

Crop Control

Flower Morphology and Fruiting Physiology

A s with all fruit crops, peach fruit forms as the result of the pollination of flowers. Peach trees bear "perfect" flowers, meaning that
when fully and normally differentiated, the flowers have both male
and female parts. The male portion of the flower is known as the
stamen. It is made up of anthers, which produce and shed pollen, and
filaments, which support and hold the anthers above the female part
of the flower. Pollen is shed both before and after flowers open. Most
peach flowers are self-pollinated; insect activity can improve fruit set
but is not normally needed to set a full commercial crop.

The female part of the flower is known as the pistil. It is composed
of a stigma, the ovary, and the style, which connects the two. After pollen is shed the fertilization process begins when a pollen grain lands
on the stigmatic surface and germinates. It produces a pollen tube,
which grows down the style and penetrates and fertilizes the embryo,
which is housed in the ovary. The embryo (seed) continues to develop
as the fruit matures and is the source of plant growth regulators that
are important in increasing fruit size. Without viable seeds, fruit either abort or remain as small, unfertilized fruit, which may or may not
ripen. Even if they do remain and ripen, these fruit (known as dinks,
runts, or second crop) will be small, will not develop normally, and are
of little commercial value.

Flowers are initiated in summer. In very stressful summers, heat

119

Parts of a peach flower include both male and female components.

or drought stress may cause abnormalities in the development of the flower. In some cases more than one pistil can be initiated, and multiple fruit can be generated from the same flower. These fruit are joined together and are called twins or doubles. Such fruit are normally just two, but there may be as many as six or seven individual fruit that are joined together and inseparable. While they are somewhat amusing, these twins are not salable in the wholesale market and are usually avoided by consumers in a retail setting. They can, however, be a serious economic problem during the thinning process. Selectively removing twins and leaving normal single fruit can be extremely time-consuming and costly.

Thinning

To the new grower, thinning the first crop is a practice that is emotionally difficult. After working for two to three years to grow a vigorous tree and produce flowers, knocking most of the fruit set to the ground in the first year of production seems at best counterproductive and at worst a crime against nature. It is important to understand that for peach trees, just as any for other perennial tree, the purpose of flower-

Heat and drought stress during summer flower initiation may give rise to flowers with multiple ovaries that can set multiple fruit.

ing and fruit set is to produce viable seed. Agriculture is a manipulation of this natural process in which we seek to produce fruit for human needs. In many years, a mature tree produces 8,000 to 10,000 flowers, but the tree can only size and ripen 600 to 800 fruit without severely overcropping the tree. Thinning fruit reduces the photosynthetic demand on the tree, allows the edible portion of the remaining fruit to increase in size, and allows the tree to produce a healthy canopy of leaves necessary for ripening the current year's crop. Because peaches are produced only on one-year-old growth, balancing fruit load with vegetative vigor is important in producing a good crop potential on an annual basis.

On a well-thinned tree, the size and weight of a single large peach can be greater than those of several peaches on an under-thinned tree.

It is important to maximize fruit size in order to obtain the highest price for the crop. In many cases the price for a half-bushel box of 2.5-inch peaches is twice as much as for 2 -inch-diameter fruit. It is also much easier to find a buyer for large fruit.

Thinning is extremely labor intensive and must be done in a timely fashion in order to obtain the desired increase in size and quality for the fruit that remain in the tree.

When to Thin

The earlier peaches are thinned, the greater the response in increased fruit size and quality. The greatest response is obtained when the blooms are thinned, but bloom thinning is suggested only on very early maturing varieties, which are typically relatively small in size. Bloom thinning is expensive and is suspected to increase the number of fruit with split pits. While early thinning results in greater fruit size, it is important to understand that the earlier an orchard is thinned, the greater the risk of economic loss if freezes or frosts are encountered. For example, a frost that kills 90 percent of the fruit on a thinned tree may reduce the crop by 90 percent, while an unthinned tree may still produce a full crop following the same weather event.

The peach fruit develops in three basic stages. During the first twenty-one to thirty days after bloom, the fruit increases in size rapidly due to cell division. During this time the peach pit gains most of its size and begins to mature a viable seed. Beginning late in the third week or early in the fourth week, the fruit sizes more slowly and appears to be just hanging on the tree. This slow fruit development stage lasts through the fourth, fifth, and sixth weeks after bloom. At this time the peaches are a little larger than your thumb and the pit is soft. For logistical and practical purposes, it is recommended that peach fruit be thinned during the fourth, fifth, and sixth week after bloom.

Pit hardening begins about seven to eight weeks after bloom and signals the end of cell division within peach fruit. Thinning fruit after that time usually results in no significant increase in potential fruit size. Earlier varieties need to be thinned first and the later-maturing varieties last.

Fruit Growth & Development

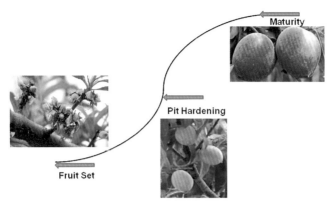

This sigmoid curve shows the progression of fruit size from fruit set to maturity. The first half represents cell division, and the second involves cell enlargement within the peach.

While thinning after pit hardening will not result in an increase in fruit size or fruit quality, it may reduce the stress on the tree. Ripening fruit demands a tremendous amount of carbohydrate energy, and an overcropped tree will suffer from reduced shoot growth. Remember, in any year, you are growing two crops: the fruit that will be produced that year and the wood that will be responsible for the following year's crop. Overcropping trees can and usually does result in a tree that will bear much less fruit the following year as well as predisposing the tree to winter injury. In some cases the sheer weight of the fruit on an overcropped tree can cause the limbs or trunk to break or split. Trees suffering this type of damage rarely recover.

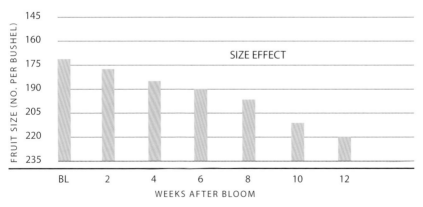

The effect of fruit thinning on 'Redhaven' fruit size at two-week intervals.

An over-cropped tree can break or split under the weight of the fruit.

How Much to Thin

In general, growers tend to underthin producing orchards. A practical rule of thumb is that peaches should be thinned to 6 to 8 inches apart. This means not just 6 to 8 inches apart on each limb but 6 to 8 inches apart in every dimension among limbs. It is important to focus on how many peaches remain on a tree rather than how many have been removed. Under Texas conditions, the *average* yield on a producing tree is four half-bushel boxes per tree. With optimal conditions, properly trained, well-thinned trees can bear twice this amount. Growers should remember that there needs to be a balance between canopy and fruit. If there is an insufficient canopy, fruit will be of small size and poor quality, and the tree will suffer. Thinned fruit are normally just left where they fall. While on rare occasions fungal pathogens may infect thinned fruit, it is economically impractical to remove this fruit unless you only have a few trees.

Further supporting the practice of fruit thinning, the accompanying tables show the direct correlation of fruit size and yield. These data show that mature trees need to carry from 600 to 700 peaches in order to produce 350 bushels per acre if the fruit diameter is 2.5–2.75 inches.

The rule of thumb is that fruit should be no closer than 6 inches apart in every direction on the tree. The small fruit in hand have been removed to attain 6-inch spacing between remaining fruit.

When thinning peach trees, it is important to focus on how much fruit remains on the tree, not how much is on the ground.

How to Thin

At present most growers still rely on hand labor to thin peach trees. Cushioned broomsticks, bats, or poles are commonly used to thrash trees gently so as to remove the majority of the unwanted fruit. Hand

Average weight in ounces and number of peaches per half bushel according to size.

Size (inches)	Weight (ounces)	No. of Peaches per Half Bushel
1¾	2.30	191
2	3.00	146
2¼	4.10	107
2½	5.55	79

Number of peaches per tree required to produce eight half bushels of peaches.

Size (inches)	Number of Peaches
1¾	1,532
2	1,172
2¼	860
2½	636

follow-up consists of breaking up clusters of fruit and fine-tuning the number and placement of the remaining desirable fruit. With the diminishing supply and increasing cost of farm labor, growers are experimenting with mechanical and chemical methods to reduce the labor demand.

Mechanical shakers have been used in the past, but in some cases shaker design or operator error has caused severe trunk or limb damage. Mechanical thinning necessitates waiting until fruit are large enough to be removed when shaken, resulting in less than optimal fruit size. Nevertheless, in the past few years skilled custom operators have made major contributions in some orchards where heavy fruit set has overwhelmed available labor sources. Varieties differ in the amount of force required to remove fruit, whether by hand or machine thinning. With mechanical thinning, fruit must be about the size of a quarter to have enough mass to fall from the tree when shaken. Tree training is a vital part of developing orchards for mechanization of fruit thinning. Trees need to be developed with trunks 20 to 24 inches high and with scaffolds radiating upward and out. Downward-hanging limbs and hangers should be removed. Corrective pruning in some of the older orchards would prove beneficial.

Numerous chemical agents designed to abort flowers or fruit have been evaluated as fruit thinners over the past thirty years. At this point there are no chemicals labeled for this purpose. At present the most promising compound appears to be ammonium thiosulfate, a fertilizer that kills open flowers during full bloom. Rates, timing, and differences in varietal response need to be evaluated before this product can be recommended reliably. It is important to note that even though it is a fertilizer, if recommended and used as a plant growth regulator to thin fruit, this compound must be registered with the Environmental Protection Agency as a pesticide. It is doubtful that any producer will be willing to label this product as a thinning agent because of product liability issues.

Some peach varieties, such as 'Junegold,' are prone to a physiological disorder called split-pit. This occurs when the fruit enlarges rapidly and creates a hollow cavity within or around the seed and splits open at the stem end of the fruit. When this occurs, fungal decay organisms can invade the fruit and colonize seed and fruit tissue. Some of these fungal contaminants are known to be strong carcinogens, making the whole fruit unfit for consumption. Wholesale markets are entirely intolerant of split-pit fruit, and retailers should strongly consider not marketing this fruit for health and liability reasons alone. Sources differ on how to prevent split pits; while some sources suggest very early thinning, others suggest thinning close to pit hardening to avoid a large increase in fruit size. Growers should avoid varieties with this trait and experiment with the timing of thinning when it does occur.

Some varieties are prone to the disorder known as split-pit.

Chapter 13

· ·

Harvesting and Handling

The harvest of a peach crop is the culmination of years of work and many ups and downs. Weather extremes have done nothing but make the experience worse. Just when you thought you had it figured out, the temperatures dropped to 25°F on April 8 or rocks fell out of the sky on May 18, which either reduced your crop to nothing or made it very hard to sell. Harvest can be a bittersweet ending to a job best described as *hard work*. Depending on the size of your operation, whether one tree or several hundred, you need to know what you are going to do with the fruit and move it to its final destination, be it sell it or smell it!

Fruit must reach its prime maturity on the tree or the flavor will be compromised. Many have the tendency to want to harvest too early, thinking that green fruit will ripen off the tree. Peaches are a "climacteric" fruit, which means that starches can continue to ripen into sugars after the fruit is picked, but the amount of organic acids or what we perceive as flavor does not increase after harvest. If a peach is picked too green, it will not ripen properly and develop this full flavor. Firm-ripe best describes the optimum time for harvest; the stage at which the fruit have reached full flavor, yet they can still be transported.

It is common and acceptable to make multiple harvests due to the uneven ripening of peaches on the tree. Color and firmness are the two best factors to use in selecting fruit for picking. Depending on the variety and the specific weather of a season, trees need to be picked every day to every other day. A single variety will usually take ten to fourteen days from first picking to last.

Generally speaking, peaches have two color patterns. The ground color of the peach as it approaches maturity is light green. A change in color toward yellow is the first definite indication of maturity. The red overcolor is another index. The red color or blush is dull when the underlying ground color is green. When the ground color changes to yellow, the blush will be bright and easily detected. As some varieties possess more blush than others, the absence of green in the ground color is a good indicator of maturity.

Peaches are physiologically mature when the ground color changes from green to yellow.

Workers must exercise care when picking and transporting fruit.

Firmness should be used along with color to determine when a peach is firm-ripe. Peaches tend to remain firm as long as they are sizing. It is only after this final swell that the flesh begins to soften. A peach can increase by as much as a quarter inch by remaining on the tree two additional days, so one wants to make sure the peaches have reached their full genetic potential. A firm-ripe peach is best compared to the feel of a softball; there is some give but a firm background. Such peaches will hold up for several days and ripen to satisfactory dessert quality.

Harvesters must be able to distinguish between firm-ripe optimum-sized fruit, green undersized fruit, and overripe fruit. They should treat the fruit as the precious living organism it is. Excessive finger pressure and dropping the fruit into the harvesting containers cannot be tolerated. Picked fruit should be placed directly into the container used for transfer to the packing shed.

Harvest containers must be smooth, shallow, well ventilated, easy to clean, and stackable. The containers should not be overfilled, and the fruit should be handled as little as possible.

The peach is very active physiologically even once removed from the tree. Peaches have a high rate of respiration; slowing down this respiration can extend shelf life.

The best storage for peaches is at 31° to 32°F with 90 percent relative humidity. Because peach fruit contain high amounts of sugar, the freezing point is a few degrees below that of water. This allows storage at these low temperatures. Firm-ripe peaches can be stored for two weeks with minimal loss in quality. High humidity in storage slows the moisture loss from the fruit. There should also be sufficient air circulation to dissipate the CO_2 and heat of respiration. Quickly cooling the fruit can increase the storage life significantly.

Precooling the fruit by submersing it in an ice water bath rapidly removes field heat and is advisable prior to cold room storage.

Fruit harvested early in the day will contain less field heat than those harvested later. Ideally, peaches with a flesh temperature of 80°F should be precooled to 40°F before storage or shipment. Keep in mind that refrigerated trucks cannot precool a commodity; they can only be counted on to maintain temperatures.

Packing lines range from as simple as a sizing conveyor all the way to a full-blown washing, defuzzing, waxing, sorting, and sizing line complete with fungicidal treatments to reduce postharvest rots.

Unnecessary drops and rough handling of the fruit must be avoided.
Four to five months have gone into the production of this precious commodity, and it would be crazy to mishandle the fruit and destroy all that
you have worked to produce. Proper cleaning and maintenance of the
packing line will help to ensure fruit quality in the packed container.

Tree-ripened peaches are a delicacy to behold, though it is a challenge to get them to consumers in mint condition. Still, know that your efforts will be rewarded, as Texas peaches sell themselves.

Field containers must be sturdy, well ventilated, and stackable to transport fruit from the orchard to the packing shed.

Large packing sheds commonly have equipment to cool, defuzz, and size fruit prior to boxing.

Chapter 14

. . . .

Marketing

The fact that we live in a global economy is nowhere more evident than in the produce section of a grocery store. Like many other kinds of fruit, peaches are commercially grown in both the northern and southern hemispheres, resulting in the availability of fresh peaches in every month of the year. In the southern hemisphere, reduced labor costs allow for production and shipping to distant markets that would be impossible on this side of the equator. Even in North America there are a number of states where peaches are grown on a much larger scale than in Texas, where productivity is far more consistent, and where costs or economies of scale are far superior to local conditions. Often peaches can be bought in the grocery store for near or even below the cost to produce them in Texas orchards. So what do we have to offer the consumer? The answer is quality.

Peaches that are grown in distant locations must be harvested much earlier in the ripening process in order to maintain the needed firmness to arrive at the market unbruised. While their shape and color may be appealing, flavor and eating quality will be greatly inferior to those of a fruit that has remained on the tree longer and has been harvested at a more advanced stage of maturity. As noted, peaches are a "climacteric" fruit, which means that starches can continue to ripen into sugars after picking, but the amount of organic acids or what we perceive as flavor does not increase after harvest. Harvested fruit with no planned market usually results in discarded fruit or produce sold at less than ideal prices. In many ways, growing is the easy part of peach production. Every successful peach operation is a direct result of a successful marketing scheme.

Wholesale Marketing

Many growers with 40 or more acres of peaches find that they simply cannot sell all of their production in a local roadside market. Wholesale marketing means that by definition, some other person collects the retail markup when the fruit is finally sold to the consumer. While a number of growers make good profits with this marketing strategy, the wholesale marketplace can be competitive and demanding. If a buyer inspects a single box and finds any fruit that should have been culled, the entire load may be rejected. Prices are also extremely volatile, and dumping of fruit on the market from out-of-state sources may cause previously "agreed upon" prices to fall drastically.

Remember, people buy fruit with their eyes. Fruit should be large, well colored, and free of blemishes. The investment in timely fruit thinning will more than pay for itself in increased prices. Fruit must

▲ Larger wholesale growers commonly have their own packing sheds with processing and storage facilities.

◄ Wholesale growers typically pack fruit into 25-pound half bushel boxes.

be harvested while quite firm in order to be shipped, displayed, and sold before it becomes overly mature and soft. Growers usually wait until the background color starts to change from green to yellow as a gauge of maturity. Most varieties, if well exposed to sunlight, develop a pink or reddish blush on about half the surface of the peach. This effect of light exposure is another reason that canopy management is so very important. Fruit are typically picked into field lugs, then sorted, graded, and packaged into 25-pound half-bushel boxes. Many wholesale operations use automated grading tables that commonly have sizing, washing, and defuzzing operations. Choosing boxes that can be stacked and that will hold up to moisture condensation in cold storage is critical to continued successful marketing.

Farmers' Markets

Urban residents are increasingly recognizing the advantages of locally grown produce, from both quality and world impact perspectives, and more important, people are willing to pay for this. Because of this trend, many medium to large cities have organized farmers' markets to allow for local sale of produce directly to the consumer. Farmers' markets are commonly governed jointly by the municipality and a board of directors, usually made up of farmers and consumers. Each market has its own rules, but generally they restrict sales to growers within a given distance from the market and require that vendors grow what they sell. Markets may also limit the number of growers who offer a specific crop. The problem for peach growers is that peach trees are usually harvested every other day or so, and the fruit have a relatively short shelf life. A given farmers' market usually operates only one day a week. Hence to sell an entire crop successfully, a grower may need a presence at more than one market, ideally on different days. For some growers, farmers' markets offer a way to supplement roadside sales, which are usually more brisk at weekends than on weekdays.

Roadside Sales

The demographics of the peach industry are changing, and increasingly growers are choosing to retail fruit directly to the public from permanent roadside stands. Traditional peach-growing areas such as the Hill Country are known for both production and retail stands operated by area growers. Rather than viewing one another as competitors, growers have come to recognize that the concentration of market stands represents a huge drawing card for consumers from nearby Austin and San Antonio. Other areas of the state are also known for their local production and continue to draw customers both locally and from urban areas. Collective advertisement pays big dividends for growers who are in close proximity to one another. The Texas Fruit Growers' Association's website (http://www.texasfruitgrowers.org) and the Hill Country Fruit Council's site (http://www.texaspeaches.com) are both good examples of how modern technology can work for even small growers.

Fruit in roadside outlets is typically sold by volume rather than at an advertised price per pound. While customers may have no problem paying $14 for an eighth of a bushel of peaches, if they did the math and realized they were paying $2.24 per pound, they might balk at the purchase. For reasonably sized fruit, that can be about one dollar per peach. Peaches can commonly be bought for a fraction of that price in large grocery chains, but the point is that your product is vastly superior in quality and worth the price premium. Roadside stands also offer peach growers the opportunity to market other crops they may grow. In order to reduce the risk of economic losses, some growers have diversified their operations and grow other fruit or vegetable crops. Growers have learned that having peaches is a great way to draw people into the stand, and while there, consumers are apt to buy other farm-fresh produce as well.

People buy fruit with their eyes. Attractive fruit and displays are important aspects of retail sales.

Pick-your-own marketing has tremendous benefits and also a number of drawbacks, making it both attractive and challenging. The elimination of a labor crew needed for harvest is perhaps the greatest benefit to this marketing technique, but inhospitable weather can keep patrons away, resulting in a crop that becomes overripe on the trees. Fruit in pick-your-own operations is typically sold by the pound. For varieties that may be low yielding but are in high demand by consumers (like 'Loring'), prices may be higher than for fruit that ripens at a different time of the year. Many patrons will pick and purchase fruit that may be overripe or small or have a cosmetic blemish, when that fruit would otherwise be unsalable in any other setting. For many customers, it is not merely the produce but the whole experience of going to an orchard that makes the trip worthwhile.

Retail markets can range from storefronts and permanent sheds to small temporary seasonal stands.

Pick-your-own orchards have many positive benefits but are dependent on good weather for family outings.

To survive a pick-your-own operation, a grower must like people. Visitors will want to know what it is like to grow fruit, so be prepared to answer the same set of questions day after day. As with any setting, the vast majority of people who come to your orchard will be courteous, friendly, and happy to be on an outing in your orchard. Others will climb in your trees and break limbs, discard perfectly good fruit, argue with you about prices, and just be plain unpleasant. You must be able to let those experiences roll off your back and not affect your mood or your optimism. A grumpy setting is not conducive to repeat customers.

Successful pick-your-own operations are well manicured and clean. Be prepared to give verbal direction as well as providing signage for where patrons should be picking, and it is helpful to have a staff person in the field directing people and helping them learn which fruit is ready to be harvested. Restroom and washing facilities and a shaded area with benches are needed to help patrons who may not be accustomed to the heat or the exercise. Consider having patrons sign a guestbook and leave mailing and email addresses. These lists can be valuable assets for advertising in future years.

. . . .

Financial Needs to Get into the Peach Business

P
each production is one of the most intensive forms of agricultural production. Establishment of a peach orchard is both labor and capital intensive and should be attempted only after a complete economic analysis of the cost/benefit relationship. A considerable investment is required to establish the orchard and purchase the necessary machinery and equipment before the first harvest of peaches can be expected. Labor requirements of an intensively managed orchard must also be recognized. Negative cumulative cash flows can be expected for five to eleven years, or more, depending on production, prices, and costs of production. As a result, financial needs to get into a fruit orchard business must be carefully considered.

Although a 20-acre model is used in this chapter, the costs are discussed on a per acre basis. Equipment costs shown in table 15.1 are on a total cost basis. Equipment investment per acre is high, and economies of scale considerations are an important planning factor. The same amount of equipment is often needed for a small orchard as for a larger one. Thus the equipment shown in table 15.1 could be used to support a larger orchard, thereby reducing the per acre overhead cost. Income tax aspects of the investment are another essential planning factor, but individual circumstances vary so much as to be beyond the scope of this book.

Initial Investment

Investment is the capital an owner/grower has invested to establish the "manufacturing plant," in this case the orchard. In planning the establishment of an orchard, a timetable of investments to establish and conduct an intensively managed operation is essential to allow a complete economic analysis of the investment.

Capital investment during the first year, excluding land, approximates $77,806, as shown in the following table. The investment shown in this table is the minimum amount required to operate an intensive, well-managed peach orchard. Items shown are assumed to be new. Good used equipment is available and could be used to reduce initial investment requirements. Figures may have to be adjusted for inflation, but the relationships are the same.

Overhead Cost

Although the expenditures to purchase the equipment and irrigation system are substantial, they are not considered a cost but rather an investment. Investments, however, generate costs to the business. These costs are represented as the ownership cost of owning the "manu-

Equipment costs on a 20-acre orchard (2010 dollars).

Equipment Investment		
Tractor 55 H.P.		14,730
Pickup (orchard portion)		9,000
Shredder		1,576
Shed		5,000
Post hole digger		1,100
Herbicide sprayer		1,000
Air spray (PTO)		6,700
Fertilizer injector		800
Misc hand tools		400
	Subtotal	$40,306
Irrigation Investment		
Well (dig and encase 250 ft.)		15,000
Pump		2,500
Drip irrigation system		20,000
	Subtotal	$37,500
	Total	$77,806

facturing plant." Ownership cost of fixed costs, as they are normally called, includes depreciation, interest on the investment, taxes, and insurance. These costs occur annually and are unrelated to the amount of production from the orchard. The noncash ownership costs are not reflected in the cash flow projection of the operation. As a result, a cash flow analysis does not include a profitability analysis. A cash flow analysis, however, is essential to determine financial feasibility, liquidity, and the ability to service debt and meet other financial liabilities associated with operating a business.

Orchard Establishment Costs

The orchard will require approximately $1,000/acre in variable costs and $474/acre in overhead costs, for a total cost of about $1,474/acre. The variable costs include the price of the trees—about $2.50 to $3.00 a tree—and costs for weed control, irrigation, and labor to maintain the trees. The significant overhead costs are due to the high costs of the equipment and irrigation system placed into service during the first year.

Peach Orchard Development and Operational Costs

Production should begin by the third year. When production begins, variable production costs should approximate $988/acre, with overhead costs averaging about $742/acre, for a total projected cost of about $1,730/acre. Variable production costs include weed control, fertilizer, irrigation, sprays (6–10), and labor.

Production should gradually increase to about 200 bushels per acre when the orchard is fully developed. Variable costs to operate a fully developed orchard should approximate about $1,836/acre, with overhead costs averaging about $998/acre, for a total projected cost of about $2,834/acre.

Annual Cash Outflows

A year-by-year cash flow summary will demonstrate the cash flow consequences of peach orchard investments. Cash outflows associated with the machinery and equipment investments in the orchard are often treated as though the grower purchased new items and borrowed investment capital at the current interest rate to pay for the equipment over a seven- to ten-year period.

Variable costs are combined with the debt service costs to pay for the equipment and calculate the cash required per acre per year.

These total cash payments per acre per year are combined with the income generated from the sale of peaches to calculate a complete cash flow for the orchard investment. Growers can request a budget from the Texas AgriLife Extension Service Agricultural Economics Department once they have the actual figures and costs for an orchard.

Annual Cash Flows for the First Fifteen Years of Operation

Gross annual cash inflows are obtained by multiplying the annual peach production by the expected net sale price of the peaches. Peach production and sales are expected to range from none to a small amount during the first three years of establishment. Cash flow projections using the following production/price assumptions per bushel (bu) are often prepared:

Assumption #1: Production by year five @ 200 bu/acre and price @ $30/bu

Assumption #2: Production by year five @ 200 bu/acre and price @ $30/bu *but* crop failure every fourth year

The net annual cash flow is equal to gross annual cash inflow minus annual cash outflow. The peach orchard investment cash flow is negative for several years, depending on the assumption and the stage of production, while the trees move from the preproductive to the productive stage of life. Under the "most probable" conditions for assumption #1, about five years are required to offset the accumulation of negative cash flows. Usually, the cumulative cash flow turns positive in year six, but crop failures from early frost, hail, disease, and/or lower peach prices can change these projections considerably. Typically under assumption #2, where crop failures occur every fourth year, cash flow is negative until year nine or ten.

While the peach orchard can be expected to be profitable when it is producing fruit, the cash flow consequences of establishing and bringing the orchard into production need to be anticipated and carefully planned for by the grower. Peach orchard development is a capital-intensive investment, with success or failure highly dependent on the amount and quality of product produced.

Summary

Peach orchard development is an intensive investment in terms of equipment, labor, and capital. Initial investment requires about $78,000 to purchase the equipment and irrigation system needed to operate a 20-acre peach orchard. It will cost an additional $1,000/acre to establish the orchard. Although production should start by the third year, investors can expect negative cumulative cash flows for five to eleven years, or more if breakeven yields and price combinations are not achieved, before the orchard investment begins to show a positive cumulative cash flow.

Peach production is highly sensitive to freeze, hail, disease, and insect damage and a wide variety of other possible or probable problems. These problems have a direct effect on profitability and increase the risk of the investment. An attempt to develop a peach orchard should not be tried without first preparing a detailed plan, which includes financial management considerations and is based on the latest economically feasible technology available.

A fully developed, well-managed peach orchard (years four to fifteen) is projected to generate $600–$1,400 per acre profits, depending on the production level, the quality of the product, and the price received for the product. Rather than taking these numbers as gospel, use them as a starting point. Use them to develop your own numbers, but realize that these numbers are the key to your success.

Index